SDGs and Textiles

Editor-in-Chief

Hafeezullah Memon, College of Textile Science and Engineering, Zhejiang Sci-Tech University, Hangzhou, China

The book series "SDGs and Textiles" addresses the strategies to achieve sustainable development goals (SDGs) in the present, past, and future. It presents books about the present and future policies of textile ministries of different countries, and books related to sustainability education around different parts of the world in the textile sector. Moreover, it would welcome the conference proceeding related to SDGs and Textiles. The series would cover books comparing the sustainability and SDGs of different institutions and countries. The individual book volumes in the series are thematic. The goal of each book is to give readers a comprehensive overview of a different area of sustainability in the textile sector. As a collection, the series provides valuable resources to a broad audience in academia, the research community, industry, and anyone looking to expand their knowledge of SDGs and Textiles.

Textiles and life are together – life cannot be separated from textiles as it is the most important need for human beings after food. In 2015, the United Nations General Assembly proposed 17 interlinked global goals to be achieved by 2030. Since then, academia and industry have paid much attention to achieving these goals. Textile found its close relation with almost all of these 17 goals.

SDG 1 - No Poverty: Poverty would never be overcome by a charity only; it is essential to develop people's skills to have a better and wealthy life. Thus, the textile can be considered an excellent discipline to achieve this goal by creating jobs and small and medium businesses.

SDG 2 - Zero Hunger: Through the effective utilization of advanced application of Agrotech Textiles, it is possible to have higher crop yields and save crops from rough weather, unexpected rains, floods, insects, etc.; thus, geotextiles play an essential in achieving this goal of sustainable development.

SDG 3 - Good Health & well-being: There has been much health consciousness after Covid19, and medical textiles assist in getting good health and well-being.

SDG 4 - Learning & Education: Textile or fashion has remained a significant discipline for societies for ages, and there has always remained much to explore in this field. Textile-related universities may play a vital role by offering free access to their education resources, training and spreading information among the locals.

SDG 5 - Gender Equality: The textile sector is one of the industrial sectors that accepted gender equality long ago; in particular, the garment sector has more females than males. Thus, the textile sector has been doing gender equality. Moreover, there has been a recent trend for Gender Neutral Clothing, which need worth studying and may further assist gender equality.

SDG 6 - Clean Water & Sanitation: Textiles could be achieved through filtration, and of course, textile is one of the critical materials for filtration.

SDG 7 - Affordable & Clean Energy: With the recent advancement in material science and engineering, the textile sector has come on the front for, not only by using this clean energy during textile production but also by assisting the production of this clean energy, either in the form of wind turbines blades made of textile composites or by energy harvesting from T-Shirts, etc.

SDG 8 - Decent Work: Recently, there has been much attention that the textile workers are not paid well, labor rights are not cared about, etc.

SDG 9 - Industry and innovation: Textile Industry always follows innovation; the textile companies that do not chase innovation cannot survive in the market.

SDG 10 - Reduced Inequalities: Getting better life and well-being would help reduce inequalities in the textile industry.

SDG 11 - Sustainable Cities: Sustainable Textile Cities through Buildtech and transport textiles.

SDG 12 - Consumption and Production: Textile and garment consumption and production all come under.

SDG 13 - Climate Action: Oekotech or Ecotech Textile, waste management of textiles are upfront to achieve this goal of sustainable development.

SDG 14 - Life Below Water: Mitigating microfiber waste in rivers and oceans may come under the context of it. There has been much attention on this subject after passing the bill at the parliament level of the UK.

SDG 15 - Life on Land: Geotech or Geotextiles studies life on land.

SDG 16 - Peace, Justice, and Strong Institutions: Protective textiles are doing their best to achieve peace, justice, and strong institutions.

SDG 17 - Partnerships for the Goals: The application of textiles to achieve sustainable development goals is only an example. In all textiles sectors, combined efforts of all the goals are essential to achieve true sustainability.

Hafsa Jamshaid · Allah Dad
Editors

Creative Textile Industry

Past, Present and Future of South Asian Countries

Springer

Editors
Hafsa Jamshaid
National Textile University
Faisalabad, Pakistan

Allah Dad
Pakistan Institute of Fashion and Design
Lahore, Pakistan

ISSN 2948-1236 ISSN 2948-1244 (electronic)
SDGs and Textiles
ISBN 978-981-97-3801-4 ISBN 978-981-97-3802-1 (eBook)
https://doi.org/10.1007/978-981-97-3802-1

This Springer imprint is published by the registered company Springer Nature Singapore Pte Ltd.
The registered company address is: 152 Beach Road, #21-01/04 Gateway East, Singapore 189721, Singapore

Contents

Creative Industry: Introduction and Significance

Hafsa Jamshaid and Allah Dad

Abstract A short description of the literature that this work was built upon and the niche it will fill the concepts of creative textile industry with the concept of difference and its significance. Definitions, fundamentals of creative industry and different types of creative industry were discussed. Furthermore, we summarized how creative textile industry works in three countries named as Pakistan, India and Bangladesh. It has knowledge related to sectors involved in the manufacturing of creative textile products and their examples. Need for value-added textiles in the South-Asian countries and some practical examples of creative textile products were also highlighted in the chapter.

Keywords Creative industry · Textile · Technology · Economy · Sustainability · Value addition

1 Introduction

The creative industry is a set-up of industry that involves the development of goods and their manufacturing in relevance with the original ideas, innovation and creativity. The development of goods or a service can boost the economic value through the involvement of creativity in their business line [1]. Creative industry is a business model that depends on creativity. The global marketplace undergoes continuous transformation, mirroring the evolving demands of people. Each individual aspires for change, seeking novelty and uniqueness. Effectively implementing successful changes is essential in this dynamic market [2].

H. Jamshaid
School of Engineering and Technology, National Textile University, Sheikhupura Road, Faisalabad 37640, Pakistan
e-mail: Hafsa@ntu.edu.pk

A. Dad (✉)
Pakistan Institute of Fashion and Design, Johar Town, Lahore, Pakistan
e-mail: allahdad@pifd.edu.pk

H. Jamshaid and A. Dad (eds.), *Creative Textile Industry*, SDGs and Textiles,
https://doi.org/10.1007/978-981-97-3802-1_1

Creativity is a noun that narrates the original ideas and imaginations. Any idea/object that originated from creative activity in a certain domain is a creative product or service. When industries utilized the new original ideas in the development of any product or service that leads towards the economic welfare refers to a Creative *Industry* [3]. *When the sense of imagination and new idea are related to a textile industry it becomes the creative textile industry.*

The creative sector is undergoing rapid transformation, with businesses in creative industries leading innovation and shaping our lifestyles. Streaming services, online shopping, mobile apps, and doorstep deliveries—all owe their existence to the creative industries. The creative industries included the industries that combine of creation, production and commercializing of creative content, which can be goods or services. Situated at the intersection of arts, culture, business, and technology, the creative industries thrive on intensive creative skills, generating income through trade and intellectual property rights. This convergence makes it a dynamic force influencing various facets of our society and economy.

The creative sector is continuously progressing, emerging new job opportunities that demand fresh skills and passions. It's remarkable to think that mainstream creative roles like mobile app developers didn't exist before 2007. With ongoing technological advancements, even well-established positions, such as web developers, may require an entirely different skill set in the coming years. Creative employers seek graduates with exceptional transferable skills. They value individuals with a passion for their work and a keenness to continue learning. Success in creative companies often hinges on the ability to work collaboratively as part of a team, finding practical solutions to complex challenges, even for those who may not consider themselves inherently creative.

Creative industries have changed a lot in recent years through technology like Artificial intelligence, Virtual Reality (VR) and Augmented Reality (AR), different creative tools, etc. In order to stay in market of present era, professionals need to possess a blend of technical and artistic skills, combining with their creative vision. At present, creative industries are a combination of creativity and technology [4, 5].

The creative industry has a significant impact on economic growth and this industry is helping to boast the economy. Creative economy is one the fastest growing sectors. Creative industries include a range of economic activities. They create employment and income, promote innovation and contribute to societies for wellbeing. The world creative industries' market size was growing at faster rate and its market looks promising in the next few years. Its market size was valued at USD 2,669,224.0 million in 2022, and it is anticipated to reach USD 3,434,322.0 million by 2028. with a CAGR of 4.29% during the forecast period [6].

2 Types of Creative Industries

This dynamic sector is an umbrella term encompassing various sub-sectors, all rooted in creativity. The creative industries can be classified into different fields/industries or sub-sectors as shown in Fig. 1.

- Publishing (research and magazines)
- Advertising (boards, E boards, LEDS, campaign)
- Media (print and electronic media)
- Craft (indigenous and interventions)
- Architecture (simulation and real)
- Game industry, animation (development, characters)
- Fashion (style and trend, clothing)
- Film (web series, drama, cinema)
- IT industry, software (apps, development, research)
- Visual art (fiber, painting)
- Skin care (aesthetics, make-up/makeover)
- Textile (fiber, design, fabrics making)
- Music, performing art (dance, tunes, lyrics)

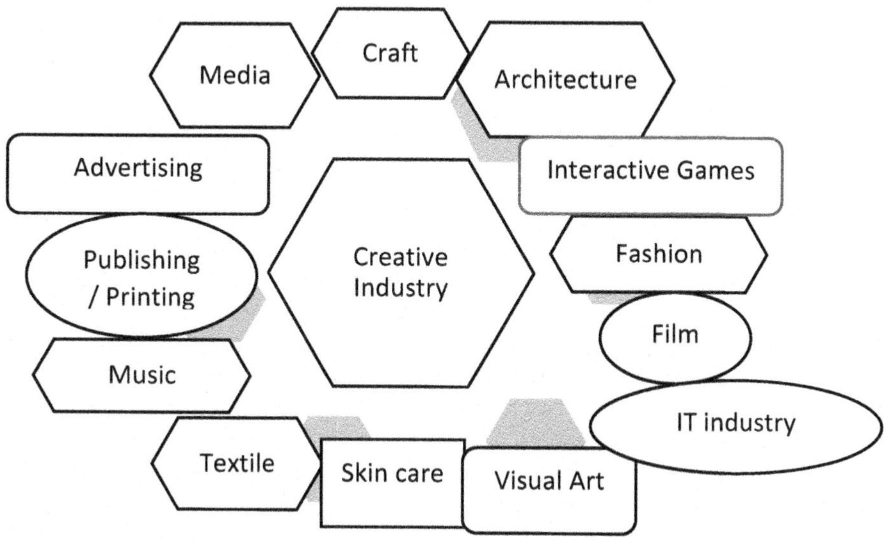

Fig. 1 Creative industry sectors

3 Significance of Creative Industry

Creative industry has multifaced advantages. The creative industry strategically situates itself at the crossroads of various crucial elements: *economics, innovation, social value and sustainability*. Economically, it plays a pivotal role by contributing significantly to the Gross Domestic Product (GDP), becoming a key driver of economic activity and prosperity. In the realm of innovation, the creative industry becomes a catalyst for growth and competition within traditional sectors. Its ability to push boundaries and introduce novel ideas fosters dynamism in various fields, encouraging adaptation and evolution. On a social level, the creative industry serves as a stimulant for knowledge and talent. It becomes a hub for the cultivation of skills, the exchange of ideas and the nurturing of artistic and intellectual prowess. This not only enriches individuals but also contributes to the broader societal fabric. Furthermore, the sustainability of the creative industry hinges on the boundless wellspring of creativity and intellectual capital. Its reliance on innovative thinking and intellectual contributions position it as a sustainable force, constantly renewing and reinventing itself.

In essence, the creative industry's strategic positioning allows it to be a multifaceted force, driving economic prosperity, fostering innovation, enriching society through the cultivation of knowledge and talent and ensuring its own sustainability by tapping into the infinite reservoir of creativity and intellectual capital.

4 Creative Textile Industry

Textile is one of the important fields and it is one of the basic necessities of human being. It has different applications like for clothing, home furnishings, functional applications. We can see textile in our daily life from towel to firefighter suit. The textile and fashion sectors are experiencing a notable transformation driven by changing consumer trends. The textile and fashion industry has social, cultural and economic significance in many societies. Textile and fashion designs are two creative industries that connect them together. Textile designers provide the innovation in fabrics to fashion designers that thrives on innovative and aesthetics. Creativity is a must in the field of textile and fashion design. It helps designer to stand out in the competitive industry and capture the attention of consumers. It also plays its roles in pushing boundaries and settings new trends in the fashion world. The creative textile industry involves various sectors that involve the use of textile in artistic and innovative way. It includes often combination of traditional textile techniques with contemporary design and technology. Some key sectors are textile art, fashion textiles, costume design, interior design/textile, technical and smart textiles, textile and surface pattern designs, textile manufacturing and innovation and textile recycling and sustainable design. The creative textile industry is an inventive and vibrant sector that syndicates customary artistry with modern design. It incorporates a variety of deeds such as textiles, fashion, home furnishings and accessories and shows business a dynamic role in the inclusive economy [7–10].

This industry is significant for several reasons:

1. *Economic involvement:* The creative textile industry is one of the leading businesses that exists worldwide, offering the jobs for masses of individuals. It creates considerable income and adds to GDP progression in numerous states. In adding, it nurtures entrepreneurship and chains small and medium-sized enterprises, encouraging economic development [11].
2. *Cultural conservation:* Creative textile construction often comprises customary weaving techniques, dyeing processes and embroidery. These crafts are handed down from end-to-end compeers, preservative the cultural legacy and inhibiting customary abilities from being vanished. The productiveness aids put up with cultural multiplicity and encourages the indebtedness of altered artistic customs [12].
3. *Sustainability and innovation*: As Ecological Fear cultivate in current era; the creative textile industry is implementing sustainable does. Sustainability and creativity can go hand in hand in the textile industry. As designers and manufacturers strive to reduce the environmental impact of textile production, they are also exploring creative and innovative approaches to sustainable design [13]. Designers and producers are progressively by means of eco-friendly constituents, embracing ethical manufacturing approaches and see the sights of recycling and upcycling practices. Furthermore, the industry is frequently asserting the precincts through technological progressions, digital printing, and smart textiles, driving modernization crosswise the sector.
4. *Fashion and design influence*: The creative textile industry is diligently knotted to the fashion and design areas, impelling trends and customer selections. Textile designs and arrangements enlighten the latest fashion works, interior design and aesthetic selections. The industry turns as a catalytic agent for originality and stimulation in numerous arenas, modeling visual culture and the way we observe and relate with our settings.
5. *Cultural interchange and global trade*: The creative textile industry assists cultural interchange and global trade. Textile designs, styles and materials are frequently inclined by many cultures and travel all over the place in the world, stimulating intercultural know-how and appreciation. Furthermore, the industry plays a significant role in global trade conglomerates, nurturing economic interactions and helping the international marketplaces.

5 Creative Textile Industry in South Asian Countries

The creative textile industry in Pakistan, India and Bangladesh shares similarities in terms of traditional craftsmanship, cultural influences and a focus on textile-based arts. Each country has its unique strengths and specializations within the creative textile sector. For Pakistan, different types of traditional printing (Batik printing,

Fig. 2 Different products of creative textile (authors own)

Ajrak printing), traditional embroideries of different regions like Balochi, Sindhi, Phulkari, etc., traditional outfit, i.e. Shalwar Kameez. For India, traditional dresses like Sari, Lehenga and Anarkali, embroideries like Chikankari (Lucknowi Embroidery), Zardozi (use of metallic thread and beads) and, for Bangladesh, Jamdani Sari. Bangladesh is also for jute-based products like rugs, etc. Figure 2 shows different products.

Role and significance on country's economy (Pakistan, India and Bangladesh) are shown in Fig. 3 and Table 1

Discussion of the Main Determinants

1. Employment Generation: The creative textile industry has created a substantial number of job opportunities in Pakistan, India and Bangladesh. From skilled artisans to textile designers, the industry has provided employment to many people, particularly women in rural areas. This has helped reduce unemployment rates and improve the standard of living for many individuals. The employment in these countries involves in fiber division, yarn making, fabric manufacturing, fabric dyeing and finishes, garment and stitching and in retail shops. The Creative industry helped in the development of the brand acroos south asia such as in India, Pakistan and Bangladesh by notable names such as Kahddi, Fabinda, Biba, Manish Malhotra, Lime Light, Sana Safinaz and Cat Eyes etc. Cat's Eye is a leading clothing brand in Bangladesh and Yellow.
2. Foreign Exchange Earnings: The creative textile industry has been a major contributor to Pakistan's, India and Bangladesh foreign exchange earnings. The country exports a wide range of textile products, including traditional hand-woven fabrics, embroideries and modern fashion garments. These exports generate significant foreign exchange, helping to stabilize the economy and improve the balance of payments [11]. Through export industry related to textile and fashion, a substantial amount of foreign exchange reserve is earning by these countries.
3. Value Addition: The industry adds value to Pakistan's, India and Bangladesh raw textile materials by transforming them into finished products with unique designs and aesthetics. This value addition helps increase the price of the products, leading

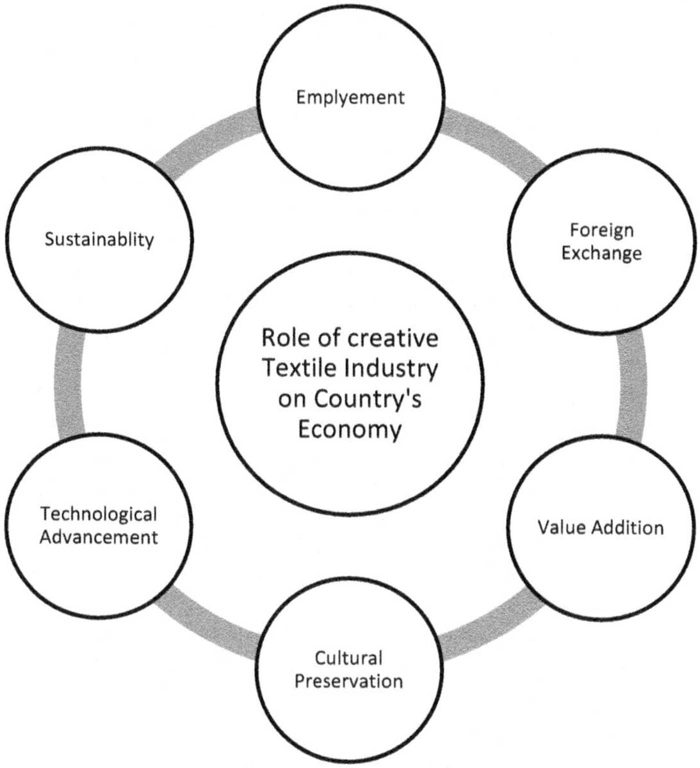

Fig. 3 Significance of creative industry on economy

to higher export earnings and overall economic growth. The supply and value line chain of textile and fashion industry is responsible in value creation through the products. For example, a bridal dress made in Pakistan involves design development arrangements of fabrics, dyeing, embroidery, hand sequence works, pattern making, stitching, etc., whereas in every step, the value is added and helps in maintaining the concept of value addition.

4. Cultural Preservation: Pakistan's, India's and Bangladesh's creative textile industry plays a crucial role in preserving the country's rich cultural heritage. Artisanal techniques, traditional motifs and indigenous crafts are incorporated into modern textile designs, which helps preserve and promote Pakistani culture on a global scale. This, in turn, boosts tourism and cultural exchange, contributing to the overall economy [15]. The brand like Sabya Sachi is good example in cultural preservation of Indian crafts and lifestyle through their motifs and designs. Similarly, the clothing brand of Pakistan Rang Ja is also an example where the design, colors and embroidery like running stitch, gotta work, cut lines and motifs represent the concept of cultural preservation. Furthermore, the brand like Aarong

Table 1 Statistics w.r.t creative industry determinants (Pakistan, India and Bangladesh)

Country	Employment	Foreign exchange	Value addition	Cultural preservation	Technological advancement	Sustainability
India	20%	44.4Bn $	Raw cotton to suiting fabrics	Yes, by Cotton Muslins, Jamdani, Tapestry	Hand spinning to compact spinning	Certified green products, use of vegetable fibers like banana fiber
Bangladesh	40%	42.6Bn $	Raw jute to tote bags, cotton fabric to garment	Yes, like Jamdani, Garbhasuti or Asmani	Hand jacquard to electronic jacquard or	Certified green products like the use of organic cotton in shirting cloth
Pakistan	40%	20.0Bn $	Raw cotton fiber to denim pant	Yes by Khaddar, Lawn fabric, Chunri	Normal dyeing to nanobubble or spray dyeing	Certified green products like use of hemp or lyocell in denim-making

Source Niazi, Muhammad Abdullah Khan, 2019: Report on Determinants of Export Performance: A Comparative Study of Pakistani, Indian and Bangladeshi Textile Sectors [14]

from Bangladesh also depicts the same concept through their design intervention based on regional craft Jamdani.

5. Technological Advancement: The creative textile industry has encouraged innovation and technological advancement in Pakistan, India and Bangladesh. Manufacturers and designers have embraced modern tools and technologies to improve production processes, quality control and product development. This has not only enhanced the competitiveness of country's textiles in the global market but has also contributed to the overall technological progress of the country. The usage of Artifical intelligence cloud computing, block chain data management the brands like LAAM from Pakistan serve the one of the best customer services. The advance level of equipment like robots in textile industry boosts the production level and quality assurance.
6. Sustainable Development: The creative textile industry in these three countries is increasingly emphasizing sustainability and eco-friendly practices. Many manufacturers are adopting organic, fair trade and environmentally friendly techniques in their production processes. This promotes sustainable development, protects the environment, and contributes to country's global image as a responsible textile producer. The utilization of the material like hemp, vegan leathers, organic cotton and vegetable fibers in their products promotes the sustainability and eco-friendly environments. The technology like nano-bubble technology in dyeing and smart foam technology in washing of garment also depicts the sustainable development in these three countries.

6 Need of Value Addition

With the rise of globalization in recent years, competition has intensified, and consumers have become more quality conscious. Technological advancements and evolving standards have elevated consumer expectations, prompting producers to enhance their goods to meet these heightened standards. In the contemporary global landscape, lifestyle products exported to various countries encompass value-added items. This phenomenon is driven by shifts in consumer preferences and trends. The world marketplace is continuously changing and so is the demand of people changing. Every person desire for some change, i.e. something new and unique. The successful effective implementation of change has to be done to in the market. Recognizing this, it is imperative for the textile industry to delve into evolving designs, patterns, product development and necessary adjustments in production facilities. This involves diversifying materials, production techniques and related expertise to secure a leading position amidst the intensifying competition with other nations. The current reality underscores that to withstand fierce competition, adopting value addition techniques is crucial echoing the principle of "survival of the fittest."

In recent years, there has been a growing emphasis on sustainability in the textile industry, driven by environmental concerns and consumer demand for eco-friendly products. Several innovative approaches have emerged to address the environmental impact of textile production and consumption. Some notable innovations in sustainability in textile are using material innovations in respect of fibers and natural dyes, recycled and upcycled materials, biodegradable fibers, closed-loop manufacturing, water-less dyeing techniques, nanotechnology, 3D knitting and printing, etc. The purpose of all these is reducing the load on resources and minimization of waste. These innovations collectively contribute to making the textile industry more sustainable and environmentally friendly. As technology continues to advance and consumer awareness grows, the industry is likely to see further developments in sustainable practices.

The production of high-value-added apparel products is crucial for the sector's survival. Value addition in textiles can significantly benefit developing countries like South Asian countries in several ways. Developing countries can generate more revenue from their textile products, by diversification of products leading to increased income and economic growth. Value addition allows for the production of a wider range of textile products. This can also create more employment opportunities in the country.

7 Disadvantages of Creative Industry

Some of the disadvantages of the creative textile industry are given in Fig. 4. In this figure, the author tries to depict all those factors that are mainly responsible for creating cons of creative industry. The competition is very high in this sector because all intellectual brains are working for their survival and betterment, therefore, the competitiveness is very high let's take an example of leading two soft drinks, PEPSI and COCACOLA, both belong from creative industry of beverages and their competition is very tough in terms of market share and market growth. Also, the thought process involved in developing ideas is very out of the box, which creates the problems in terms of its formulation, the ideas become very far from the reality, for example the textile industry made an idea to develop the collection with 100% recycle cotton yarn, which becomes difficult to convert into reality because of strength and functionality.

Similarly, the Intellectual Property Management also becomes a problem in terms of getting rights as patent, copyright at the time of registration. The heads of becoming a copyright or a patent content sometimes needed to be reevaluated. The compliance or audit para also comes and creates an issue due to less common understanding of submitted and approved titles of certificates. The procedure of registration becomes very lengthy especially when it comes to third-world country due to availability of less technology and research. The time-consuming procedure, most of the times, became responsible of losing the first mover advantage in third-world countries.

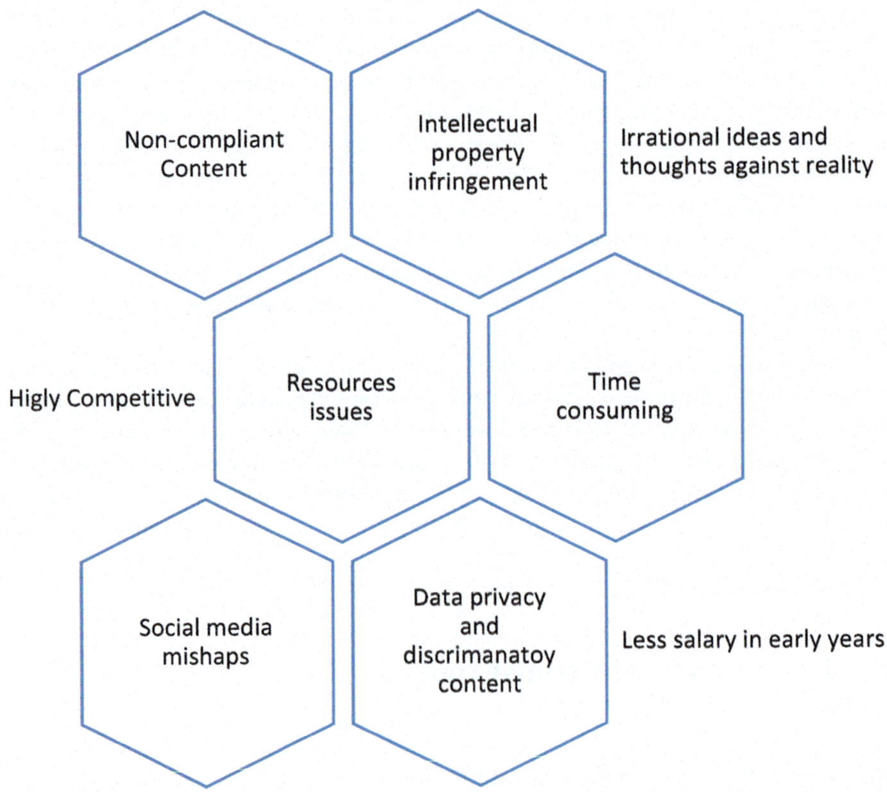

Fig. 4 Disadvantages of creative industry

Furthermore, in creative industries, the business model became very difficult to gets it take off position, which leads towards difficult to retain the human resource due to less wages or salaries in the beginning. For this, the scarcity in the availability of skilled human resource also becomes a fundamental issue for sustaining the new idea of development. The resource materials are also not easily available in most of least and developing nations due to market dynamics, funds and production. For example there is a need of innovation in indigineous fine filament production, which is not available in Pakistan such as Viscose, Nylon etc which is causing high input cost due to imports, non availability. The material such as semi-conductor chipe also not producing in the country which also becoming scaricity of less innovation.

For any new idea from point of origin to its full adult form requires lots of time because of involvement of lots of funds, brain child and management/registration requirements. In creative industry, the research and development is also considered a back cone of every ideation to materialization, which takes lots of time and cost, for example, development of new yarn from natural plants such as the hyacinth plant requires lots of experiments for their proper functionality and requirements.

The data management in terms of discriminatory content and privacy matters is also one of the major cons of creative industry due to availability of data similar to that and also sometime chances of being stolen in the creative industry. The competitor of any sector always keeps an eye on new ideas made by their competitors. To create social media content for the campaign and marketing becomes difficult for common understanding and other socio-cultural factors of the regions. The involvement of technology and its proper utilization in social media or advertising the content requires proper target audience, target market or a target in order to avoid any mishaps.

8 Examples of Creative Industry

To use wasted selvedge as a whole by using minimum amount of energy and processes. One of the ways in which sustainability and creativity intersect in the context of textiles is shown in Fig. 5.

A home textile work is made by a textile designer, Ms. Wajeeha Akram. Her work is shown in Fig. 6a, b and c. Her work is based on theme-based design development for a home textile collection. She planned fabric quality by utilizing relevant yarn, fabric, weaving, printing and stitching and displayed by incorporating the visual merchandising.

Let's take another example of a textile designer, who converted simple calico fabric into interior panel by incorporating the history of Heer Ranjha with their characteristics and symbols through painting and embroidery. Her work is shown in Fig. 7

The designer not only developed interior panels but also tarnsform into running yardage of textiles by following the same theme and concepts as shown in Fig. 8.

Fig. 5 Home furnishings by textile waste (authors own)

Another example is given in Fig. 9 from the apparel creative industry, here work of a designer Ms. Fatyma Amanat who worked from the concept of the latest trends by incorporating the textile fabrics (denims, trims and accessories) together to bring a product for fashion world. Her work selected by a denim industry named as Naveena Denim, which is presented in front of international buyers.

9 Conclusions

To thrive in today's market, a product must go beyond mere quality and incorporate additional features. This shift has brought the market to a point where manufacturers face a considerable challenge in promoting their products unless they offer consumers something novel. This not only rewards the manufacturer for their innovative ideas but also entices consumers with the prospect of significant profit gains. The pursuit of innovation, driven by the desire to boost profits, capture a larger market share or cater to trends, can be described as adding value to textiles. In this contemporary landscape, creating something innovative has become essential for manufacturers to distinguish their products and stay competitive in an environment where consumer expectations continuously evolve. Imagination, innovation and technology take textiles to new horizon. Value-added products are need an hour. By integrating sustainability into the creative process, designers can contribute to a more responsible and eco-friendly future for the textile industry. Creative solutions that prioritize both aesthetics and sustainability are crucial for fostering positive change in consumer behavior and industry practices. By concentrating on value-added finished products, the ambition of attaining global leadership in textiles and apparels can be achieved.

(a)

(b)

(c)

Fig. 6 **a** Home textile products by Ms. Wajeeha Akram (permission attached)

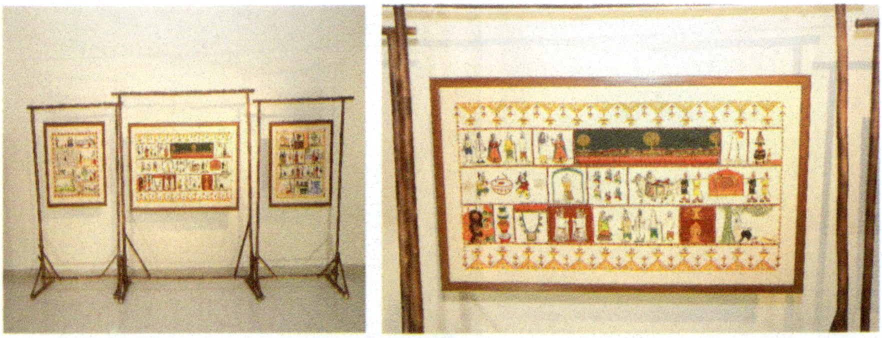

Fig. 7 Interior panels by (Ms. Rida Textile Designer)

Fig. 8 Running yardages by Ms. Rida Textile Designer

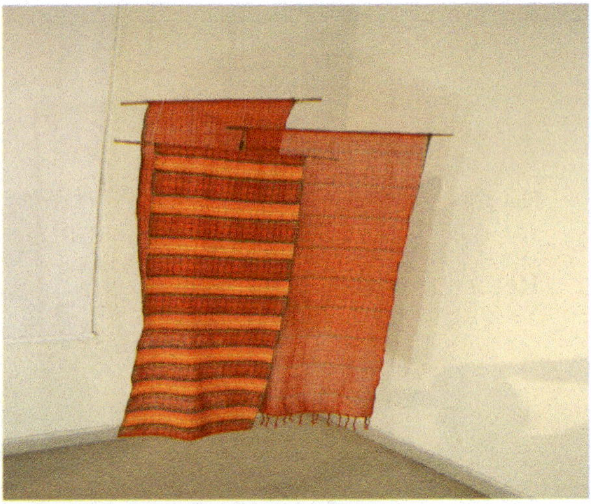

Fig. 9 Fashion apparel by Fatyma Amanat (permission attached)

References

1. Shaban A, Vermeylen F, Handke C (eds) (2022) Creative industries in India. Taylor & Francis Group
2. Boccardelli P (2016) Creative industries. In: Augier M, Teece D (eds) The palgrave encyclopedia of strategic management. Palgrave Macmillan, London. https://doi.org/10.1057/978-1-349-94848-2_758-1
3. Flew T (2002) Beyond ad hocery: defining creative industries. In: Cultural sites, cultural theory, cultural policy, the second international conference on cultural policy research. https://www.marketgrowthreports.com. Accessed 12 Dec 2023
4. Martinaitytė E, Kregždaitė R (2015) The factors of creative industries development in nowadays stage. Econ Sociol 8(1):56–71. https://doi.org/10.14254/2071-789X.2015/8-1/5
5. Mylonas N, Petridou E (2018) Venture performance factors in creative industries: a sample of female entrepreneurs. Gend Manag: Int J
6. https://www.marketgrowthreports.com. Accessed 12 Dec 2023
7. Müller K, Rammer C, Trüby J (2009) The role of creative industries in industrial innovation. Innovation 11(2):148–168
8. Akhtar WH, Watanabe C, Tou Y, Neittaanmäki P (2022) A new perspective on the textile and apparel industry in the digital transformation era. Textiles 2:633–656. https://doi.org/10.3390/textiles2040037
9. Casadei P, Gilbert D (2018) Unpicking the fashion city: global perspectives on design, manufacturing and symbolic production in urban formations. In: Lazzeretti L, Vecco M (eds) Creative industries and entrepreneurship: paradigms in transition from a global perspective. Edward Elgar, Cheltenham, pp 79–100
10. Casadei P, Gilbert D (2022) Material and symbolic production of fashion in a global creative city. In: Industry's perception of the 21st century London; Creat Ind J. https://doi.org/10.1080/17510694.2022.2143664

11. Daubaraitė U, Startienė G (2015) Creative industries impact on national economy in regard to sub-sectors. Procedia Soc Behav Sci 213:129–134
12. Jeffcutt P (2000) Management and the creative industries. Stud Cult Organ Soc 6(2):123–127
13. Forst L (2022) Textile thinking in practice: creative textile design methods as research in a circular economy. In: Lockton D, Lenzi S, Hekkert P, Oak A, Sádaba J, Lloyd P (eds) DRS2022: Bilbao, 25 June - 3 July, Bilbao, Spain. https://doi.org/10.21606/drs.2022.527
14. Niazi MAK (2019) Report on determinants of export performance: a comparative study of Pakistani, Indian and Bangladeshi Textile Sectors
15. Malik K, Dudrah R (2016) Special issue on South Asian creative and cultural industries. South Asian Popul Cult 14(3):133–136

Past and Present Creative Textile Products of Pakistan, India, and Bangladesh

Asmarah Ahmed, Hafsa Jamshaid, and Allah Dad

Textile products have a rich and diverse past in the subcontinent region, with a traditional legacy that dates back to thousands of years. The region is identified for its intricate weaving techniques, vivacious colors, and beautiful embroidery work. Some of the most eminent textile crafts from the subcontinent include Ajrak, Banarasi silk, Kalamkari, Phulkari, Ralli, and Chikankari. This chapter will give a review on how textile crafts showcase the free spirit and the creative design sense of the local artisans and the evolution that marks their work till date. Moreover, there would be a brief insight into the areas where they are made and then displayed. The concept of sustainability would be examined that has been a constant struggle of the contemporary local trends and at the same time helped to hybridize these crafts in this region as a popular textile product.

The debate about the recognition for the crafts started off in the nineteenth and the twentieth century, when Greenough strived for the status of a craftsman in the perspective of industrial revolution. Arts and crafts movement that was a key drive for design recognition of that age stressed upon the fact that instead of making visual representation and its background, the key ethics to review a product should be judged according to the barometer of function, purpose, and phases of production [1]. The debate rose due to the distance felt between the very much qualified artworks and the minor occupations that is the crafts. In confronting craft, the role of the craftsman was seen in a broader perspective where he gained his position as the mediator between history, industry, and the culture. As a result, then onwards the craft's sustainability is being tried to get retrieved in many possible ways. The modern times have adapted

A. Ahmed (✉) · A. Dad
Institute of Fashion and Design Lahore, Lahore, Pakistan
e-mail: asmarah.ahmed@pifd.edu.pk

H. Jamshaid
School of Engineering and Technology, National Textile University, Sheikhupura Road, Faisalabad 37640, Pakistan

H. Jamshaid and A. Dad (eds.), *Creative Textile Industry*, SDGs and Textiles,
https://doi.org/10.1007/978-981-97-3802-1_2

various crafts with different approaches, which would be discussed further in this chapter.

Historical Significance

The earliest form of textile could be detected is string making which was accomplished by a sample of dyed and twisted flax fibers dating back to 31,000 to 36,000 years ago. Another evidence of spinning in the Neolithic time period could be traced in the form of spindle whorl artifacts, dated back to 29,000 BCE. Then onwards, textiles have been a significant fragment of human advancement, with signs of initial textile production originated in antique civilizations such as Egypt, Mesopotamia, and China [2].

However, in Indus Valley this practice of cotton farming and weaving was found on the archaeological sites of Mohenjo-Daro and Harappa. This civilization had always been known for its innovative techniques in spinning, dyeing, and weaving textiles. The cultivation of the cotton crops were ideally supported by the agricultural land near the Indus River [3]. Thus the craftsmen of these regions were curious to explore the mystery of this plant. They were inspired by the soft feel of this plant's by product cotton. After numerous experimentation they were successful to make it softer and refined. Spinning cotton with perfect skill and mastery became the epitome of local craftsmen and they further refined this expertise over the years.

In the advance of further developments in this area, Thatta, a riverbank to Indus River became a famous lively port. It flourished as a pronounced city until the great Emperor Muhammad Tughlaq who died in AD 1351. When the Indus River shifted its course, the grandeur of Thatta ended and all the commerce activity declined in this area.This craft must have an existence from the earlier days of the Summa Dynasty 1340–1520 AD in this region but its developed form was worn in the Mughal dynasty 1592–1739 AD in the royal courts. The Mughal emperors of Delhi used to rule through their governors in this area [4], which offered a diversity of cultures and its amalgamations. Since then it has been a center of craft, calligraphy, Islamic art and trade. It increased in this region in the reign of Rajputs, Arghuns, Tarkhans and the Mughal emperors. Due to such intellectual and creative activity in this area it was known as "Elderado". It had a rich craft of needlework blooming in its cottage industries, until the Britisher's came in this area in the 19th century with machine made goods. In this age the leather goods were embroidered with silk threads and exported to Europe [5].

Raw Materials and Textile Techniques

Cotton, silk, and wool are the most frequent raw materials used in the textile produce of the Pakistani, Indian, and Bangladeshi products. This region had been the suppliers of cotton to the rest of the world from centuries. Before the industrial revolution, they had been experimenting with handmade fabrics for household activities and common everyday use. This yarn or fabric is mostly derived from semi-domesticated or wild silk produced from mulberry silkworms. Sheep's wool and hair of the yak and the goat are unusually spun to mostly produce animal fiber woolen fabrics. Thus, there is a large range of materials and techniques originating from this region ranging with color manipulation in terms of block printing, hand painting, dyeing, and embroideries, involving wraps, wefts, double ikats, felts, split ply braids, flat and pile carpets, and woven textiles. Weave assembles with mainly three techniques, in general; tabby, twill, and satin [6] are commonly in use for manufacturing. Great attention to detail and ethnicity makes them more valuable. Some of the textile products famous region-wise are the following.

1 Pakistani Textile Crafts

Pakistani textile crafts have been passed down through generations and are a significant part of Pakistan's traditional legacy. They are handled with great carefulness from one generation to another in the urge to stand high amongst the crowd. It is interesting to notice how the cultural disintegration in the Pakistani society produced an amazing array of crafts. Although there were only a few of them, however, which blended well with the concept of modernist homogenization where the craft adapted itself according to the changing demands.

Textile-based crafts are one such kind of products that had been originated basically as the handcrafted artistic manifestation from the cultural roots of Pakistan where in the past it has seen its glorious days, and now it signifies the ethnicity of various regions of the country.

1.1 Small Cottage Industries

In the Mughal era, most of the crafts were made in villages and those settlements or a group of villages were the center for artisans who established themselves as units. They manually produced the hand-crafted fabric and the whole family was involved in the craftsmanship. The craft was sustained afterwards in the colonial period due to the mercantile activity in this region. The active ports made the land hub for the merchants to explore trading with Arabian areas, Southern India, and Southeastern Asia to further Europe. These trade activities brought in influences and it affected the material and technique involved in the final textile production. Further in the twentieth century when the brand shops and designers flooded the industry, hand-loomed fabrics gained another perspective, it became a cultural identity and is still in the transformation phase of becoming a product with diverse variations. However, after the Independence in 1947, there were many organizations that worked for recognizing these textile products as a major craft of the region. They were exhibited on various platforms, and it gained popularity as a craft. If we give a look at the overall scenario, forms belonging to various districts, the regional flavor is intact in their making even today. However, these small businesses run by these cottage industries have some challenges that needed to be resolved. One of the prominent ones is the lack of funding and fully equipped artisans with the updated knowledge needed to upgrade these crafts. These cottage industries could play an important role in the economy of this region if they are given proper care.

1.1.1 Zarduzi Embroidery Work

Woven textiles such as Brocade, Jamawar, Banarsi cloth are usually used to make the formal textile range in clothing or various other accessories. Initially, mostly Banarasi or Jamawar material (Fig. 1) was mostly used for Zardozi work, but now a number of materials are used for this technique. They are available in all kinds of colors and prints. These are mostly decorated with gems and Zarduzi embroidery on top. The tradition of Zarduzi is mostly linked with the royal courts of the Mughals and the Nawabs. This kind of heavy embroidery work with precious studded stones was done on Mughal costumes and afterwards it became a trend for the formal occasions like the weddings. It was revived again in the twentieth century and often used on Lehngas (traditional skirt) and Peshwaz (traditional frock) for the formal occasions [7].

Mostly it is made with metallic threads but traditionally it was made in silver pure threads or silver coated with pure silver or gold. Coiled wires and metallic discs are also done in this kind of embroidery. Mostly chain stitch or Tambour technique is used in Zarduzi work. In the Mughal period, they also encrusted gems or precious stones in these patterns on their elaborated surface designs to make them more decorative. The patterns are sometimes simple geometric patterns or they are floral motifs. Nowadays, however, simple Zari work (Fig. 2) is used in the local embroideries to lower the costs.

Fig. 1 Traditionally Banarasi cloth (Left) or silk was mostly used as a base for the Zari work. However with the changes in climate and a variety of other fabrics available chiffon (Right) and other fabrics are also used

Fig. 2 Contemporary composition of Zari work done in a floral motif (Left). Commonly used composition of Zari work motif used in the seventeenth century motifs of the Mughal period (Right). Mostly a composition of floral motifs in metallic threads along with small beaded work is done to alternate the gems and the gold threads used in the Mughal time period

1.1.2 Guta Work

Another traditional method of ornamentation mostly used on formal occasions is that of the Guta work. It has been used commonly in areas of Punjab and mostly done by women artisans. This technique became eminent due to its unique representation and bold expression. Not only is it popularly done on traditional clothes like Shalwar, Kameez, and Dupatta but also on accessories like bags, clutches, and khussas. According to the local vendors in the Ichra market in Lahore, a famous heritage city in Pakistan, the meaning of Guta is braid. Guta is basically a ribbon weaved from metal strip wefts and fiber wefts. It is a golden piece of cloth that is either used in ribbon arrangements or it is crimpled, folded, and encircled into different shapes or flowers. It is also used in zigzag or Guta ribbon embroidery. There is a large variety of ways in which it can be utilized for such needlework. It is also interesting to notice that in menswear, the men's footwear forms like khussa and Peshawari chappal are mostly ornamented with designs that are taken from the bride grooms' clothes. The motifs are mostly done in the similar manner and they are even treated in the same color range. These kinds of footwear (Fig. 3) are mostly common on Pakistani weddings and they are a necessary accessory for these functions.

Fig. 3 Guta work khussa upper, slipper style, Khussa Mehal, Lahore

1.1.3 Amli Embroidery

Amli embroidery was originally done on woolen shawls in Kashmir, also known as Kanni woven shawls. Mostly nature-inspired motifs are made in this type of needlework method. It is basically known as Kashmiri 'tanka' locally meaning a Kashmiri stitch as it is an emblem of this valley's cultural roots. Nowadays, such embroidery is also done on the upper of the bags, table cloth, and teapot covers in a variety of colors and with experimental silhouettes.

The base colors are usually like the ones used in the shawls like off-white and black or maroon (Fig. 4). It is interesting that how the artisans have used this age long technique that was traditionally only used on shawls into contemporary other forms of utility. They have hybridized it with other range of softer and more lucrative colors to suit the taste of the modern consumer.

1.1.4 Kalamkari and Sozni

Other embroidery types, popular of this region, are the Kalamkari and Sozni, Fig. 5, mostly done on high-quality pashmina shawls. According to the local workers in Azad Kashmir and the northern parts of the country, Kalamkari embroidery was introduced by the local Islamic saint Shah Hamdan in around fourteenth century. It is an interesting blend of workmanship and technique. The blending of the traditional art of Kalamkari and the technique of blockprint gives it a different appeal altogether. On the other hand, Sozni is a widely used technique for ornamentation and storytelling from then onwards in this region. It is a good example of what ornament has narrated in terms of design and character in this region through their color story and the nature-inspired motifs.

Fig. 4 Amli embroidery is mostly done in bright colours on shawls with maroon (Left) or skin (Right) backgrounds

Fig. 5 The contemporary form of both techniques are mostly used with a blend of subtle colour palette with complementary colours that bring a balance to the whole design. Sozni embroidery pattern (Left), Kalamkari embroidery work (Right)

1.1.5 Phulkari

Another type of embroidery mostly done in Punjab is called Phulkari. The basic technique of Phulkari was to connect fabrics together in patterns of flowers with the help of satin stitch and darn stitch. Initially, these patterns were made on the borders of the dupattas and chaddars (bed sheets) in Punjab but later this type of embroidery was also done on other household and apparel articles as well. The patterns are usually geometric and floral and done onto the surface using silk threads.

The literal meaning of Phul is flower in Urdu and Kari means the work. Thus, this is the technique of working with floral patterns in a specific kind of needlework. The original history of this technique of Phulkari is unknown. It is associated with the Heer Ranjha tale written by Waris Shah where Phulkari products are known to be the possession of Heer. According to a theory introduced by Flora Annie Steel in her paper written on Phulkari, that the migrant peasants from Central Asia to eastern Punjab laid the base of this technique in this region. There also might be a possibility of its link with the Persian embroidery, where a similar embroidery practice with the same meaning exists known as the Gulkari [8]. The concept behind this technique was that when a girl is born in a family, her ancestors started making a chaddar for her dowry, visualizing a Bagh (Garden). They used images of flowers, birds, lozenges in their designs in mostly bright primary colors. The tradition of Phulkari has still existed from the partition till now in parts of Hazara District, K.P.K. and Northern Punjab. It was a tradition before the War of Independence to fully embroider the whole shawl in Phulkari needlework. Typically, *Phulkari* is also done on Khaddar (woven on pit looms in relatively narrow strips). It was especially boosted by the Sikh community. They used gold and silver foss silk along with a grid in green colored running stitch. While the Rawalpindi and Chakwal regions make Phulkari patterns in red and details in green [9]. The Jatt community is specialized in the technique of Phulkari, and the colors they usually use in this embroidery done on khussa are white, red, yellow, and green. Beads or sequins are also added to make them more decorative if needed for an occasion. This technique, Fig. 6, is frequently also used on wall hangings, bags, bag packs, laptop sleeves upholstery, and wallets.

Fig. 6 Example of Phulkari done on traditional shawl (Left) and a table top (Right)

1.1.6 Ralli and Mica Work

Near to Thatta is an important center Thar, which is known for its ralli and mica (mirror) work. It is a famous practice for its workers to carry this tradition from one Meghawar generation to another. The full name of this district is Tharparkar and it covers the total area of over 170 sqkm. It is a beautiful desert with people living in ghots (a community of people) on top of hilly sand dunes. These ghots are the center of fabric weaving, embroideries, and crafts. It is interesting as to how the location of this district plays a key role in the influences it has on its crafts. As the Rajasthan desert is on the southern side connecting Thar with India and on the northern side it has its boundaries adjacent to the Cholistan desert in Punjab [10].

Although these people are deprived of many essentials, like they face the problem of storage of water, yet they have a colorful palette when it comes to the clothes and accessories they wear. Mitthi and Malahar are important centers in Thar. They have a mixed community environment, where Hindu and Muslims are living together. They share their livelihood yet maintain their distance when it comes to their rituals and customs. So, the design which is manufactured there is mostly made of colors and a hybrid interpretation of motifs (Fig. 7) that the young girls anticipate while they are making several embroidery patterns on chaddar or small patches of fabric with small mirrors in between.

According to a worker who is an agent and deals with the artisans who are involved in the making of these crafts, the most famous Jutti in central Sindh is the Turanvari khussa (Fig. 8) slippers, which are a product of these textile surface embroideries and weaved patterns. These types of khussas comprise of a narrow sole and an upper that hardly covers the toe. The upper is highly ornamented with embroideries and colorful woolen thread pompoms. These pompoms are made in bright colors of red,

Fig. 7 Tribal women in small cottage industries usually make these types of motifs inspired locally used geometric forms with bright coloured threadwork and small mirrorwork

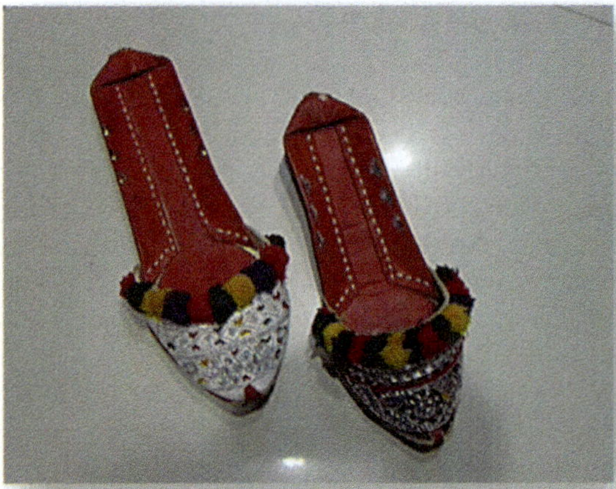

Fig. 8 Traditional Turanvari khussa with upper done in traditional mica work

blue, purple, green, and yellow. The tapering sole is often made in red with the decorative pattern of the stitch marks. However, it is very difficult to walk in these types of shoes as the sole is not very supportive, yet they are an important identity of the people of Thar.

1.1.7 Ajrak

Another traditional hub in Sindh is Hala famous for its Ajrak (Fig. 9). It is around 20 km north of Hyderabad and its foundations are laid by the Makhdum Mir Muhammad [4]. Other than the block printing of Ajrak, artisans also manufacture products like weaved blankets, wooden furniture, and lacquered tilework. One of the popular marketplaces in Hala where it is available is called the Shahi Bazaar. It is behind Sehwar Nuh. Mostly *Ajrak* is used as a symbol of pride, especially for the landlords and common people of Sindh. There is a large variety of clothing, bedsheets, tablecloth, cushions, and scarfs available in market with beautiful motifs of Ajrak. Another famous hub for these types of textile-crafted products is Jacobabad. It is a district famous for the manufacture of cotton, wooden toys, and embroideries. It was a city originated by the British establishment, and it was designed in a way that it could contribute to the trade industry. An evidence to this fact is the linking of this city to the Bolan Pass through a railway line in 1858 by British engineers [11]. Other than Jacobabad, Shikarpur had been also a vigorous mercantile center for a long time in history. Afghanis when in rule in this region had an active trade route through Bolan Pass [4]. Thus, these products were a part of the trade for a long time according to the merchants belonging from this region. Traditional bedlinens, bed covers, pouches, and shawls are a specialty of this region. Over the years, designers and local craftsmen have been experimenting with this block print technique to revitalize it with new patterns and unique motifs. It is now not only a popular craft but

Fig. 9 Vivacious colors of the Ajrak, Block printed, Sindh

also conventional form or ethnic style worn by the youth and the older generation alike.

1.2 Traditional Attires Popular in These Textile Crafts

It is interesting to see that how these textile techniques mentioned above are supported by the traditional and contemporary forms of dressing in this region. These are an important source of combining these textile products with tradition, culture, and modern adaptive trends.

1.2.1 Lungi, Kurta, Udhani, and Patka

Traditionally, the dress in the regions of Bahawalpur and the rural Punjab is that of Lungi or a Dhoti which is a draped skirt and the kurta, a loose shirt (Fig. 10). They wear it in silk or cotton garments and adorn it with different types of embroideries. The males use a turban or a Patka to support the attire with a head dress. While the female wears a Chaddar or an Udhani, which is a type of shawl to cover themselves. It is adorned with various embroideries and colors, showing the liveliness of this region.

This attire along with the khussa specifying the culture of Punjab, which is also shown in the paintings of Ustad Allah Buksh. The theme of his paintings revolved around the folklores of Punjab like Heer Ranjha. His characters of the

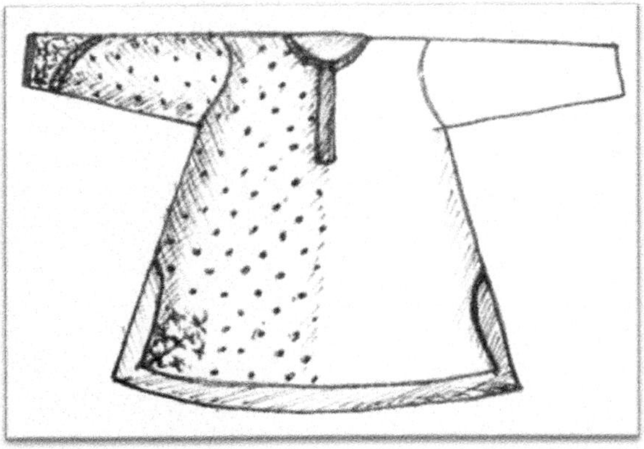

Fig. 10 Kurta

paintings showed them wearing a Lungi with a Kurta and a Saleem Shahi khussa, thus portraying a complete picture of the rural Punjab.

Nowadays, Lungi has been becoming popular on wedding occasions and mostly combined with Zardozi work and block printed techniques. Mohni Saif who had conducted many courses and trainings regarding traditional costume manufacturing and designing in institutes like TEVTA shares her observation about the potential in this market. She analyzed that new creative minds are coming in the market that are redesigning, repositioning, and making re-mockups of the already existing products like the Kurtas. She feels, the current mindset of our present generation has a large percentage of people who value traditional products of the past. For example, Zara Shah Jahan (a local Pakistani fashion designer), how she promotes and launches her products. The Kurtas she uses and generates to build an image. Even, Generations (a local Pakistani brand promoting traditionally weaved fabrics), they are all bringing back those traditional things that were once very common. So, in her opinion, there is potential in this market to contribute into the economy of our country's largest export sector.

1.2.2 Shalwar and Kameez

This is another important garment worn by men both on formal and informal occasions. A shirt made with the Boski material and little embroidery work, with a white shalwar, which are both draped garments to complete a traditional look, this type of dressing is popular in Punjab in areas like Chakwal, Gujrat, Gujranwala, Sahiwal, etc. They are also the national dress of Pakistan and are worn popularly by females too. According to Agha Hasnain Khan who has an experience in the footwear business dealings and teaches fashion retail and marketing in Pakistan Institute of Fashion and Design commented on this situation that although the people of this region are traveling and are exposed to all kind of foreign influences, still they prefer eastern wear in their clothing more rather than the western options available in the market.

Market has changed and thus the need of the hour is to change the strategy too. Now the Shalwar Kameez is being redesigned with local traditional fabrics along with the minimalistic design approach and with a blend of western cuts and patterns.

1.3 Hybridization of Design Practices

Open design forms not only produce new markets and opportunities for hybridized formations but are also means for exploiting these new challenges which are directly linked to zeitgeist and contemporary innovations in the industry. This ideology supports the contemporary notion of creative imperative where the craftsman or the designer is the sole responsible for being creative and to think out of the box with no barrier. They even engage the target audience to an extent that they can even customize the object themselves. In the twenty first century where we are existing, it is an age where there is a lot of variety and the marketing strategies have changed. The variety offered to the target group has made it more design conscious and thus need specific as well. The target group at present wants a specific product for each occasion and purpose. Such an environment is created by the flooding of brands in the market. They advertise according to every occasion and make the buyer feel the importance of their product as their need. Thus although these textile crafts are successful to an extent in redeeming themselves in the present market making it through the lens of the past but there is room for more improvisation and alterations. The question here is not only about their vitality but also how they could play a significant amount in playing a vital role in the economy of the country. However it is still commendable that small scale businesses and skillful artisans are reviving these textile crafts with experimental and affordable materials, keeping in mind the taste of the modern day consumers.

2 Indian Textile Products

Textile had been an important industry in the subcontinent region and a source of constant influence on the motifs developed throughout the history of this area. The textiles provided the base for the link from one civilization to another as there are evidence of findings of a cotton industry from the Indus civilization and then the weaving techniques mentioned in the Hindu wordings provide its continuity in the Hindu tradition too. These were indications of a booming industry to follow whose traces could be seen in the paintings of the Ajanta caves where the prince and the gods are draped in fine muslins decorated with complicated patterns. This means that the patterns evolved to such forms till the fifth century. Many Islamic expeditions from the central Asian region, Egypt and Africa started vigorous trade and as a result the manufacturers catered to their needs. Consequently, the motifs underwent a large transition and fusion with such active traffic which later was extended to parts of China and Europe. The industry saw its climax during the sixteenth and the seventeenth century when Europeans entered the Indian land during the Mughal period. The demand for Indian fabric in the European market provided a variety of block-printed designs, Cashmere, Chintz, woven woolen fabric, silk woven wraps, and muslins in the market [12].

Cotton was another durable, soft, and breathable form of clothing preferred widely in India due to its suitability for its climatic conditions. Gujarat, Maharashtra, Madhya Pradesh, Andhra Pradesh, Karnataka, and Amerali district are the famous centers for its manufacturing. A large percentage of the population is dependent on it for its livelihood. Techniques like Dabu printing, Indigo, and Ajrak are popularly done on sarees, shalwar kameez, bed sheets and pillowcases made on cotton-based woven fabrics. Colors are usually natural dyes mostly comprised of local eastern palettes.

2.1 Ikat

This is a traditional weaving technique in India, where the yarn is dyed before it is weaved into various knots and patterns. Different parts of the yarn are tied off to resist the dye resulting in a hazy misted effect when finally weaved together. This kind of technique is mostly used on sarees, scarves and upholstery. The literal meaning of Ikat is inspired by the Malay-Indonesian expression 'Mangikat', which means to wind around, bind together or simply knot [13]. Most famous motif used in this technique is that of the flame or the geometric pattern (Fig. 11). There are different types of techniques like the Single or Double Ikat, Wrap Ikat, and Weft Ikat. This was a booming industry in the past but now due to other alternatives, its craft is diminishing. It is a very sustainable textile craft with an environmentally friendly process and natural dyes used in its making [14]. The contemporary use of it on different kurtas and shirt tops are some steps to contemporize it according to the demand of present consumer.

2.2 Woolen Products

Woolen shawls around the eighteenth century were also exported to England and they were popular as Paisley shawls (Fig. 12) due to the ornament and the motif [15]. Even now woolen forms of stitched garments, woolen yarn, carpets, fabrics, blankets, knitwear, tops, scarfs, and shawls are quite famous in most of the part of India. The unique aspects are the embroidery techniques applied on them and the beautiful traditional motifs. For example, Bel is the name given to the local creepers. The pattern of Bel (Fig. 13) is usually used with small, budded flowers and delicate leaves embroidered with fine craftsmanship on locally handmade scarfs.

The local entrepreneurs and artisans have used the language of motifs to contemporize the elements into minimal yet intricate borders. They have used the motifs

Fig. 11 Commonly used geometric patterns (Left) and flame patterns (Right) on the Ikat fabrics

Fig. 12 Examples of motifs on Kashmiri embroidery done on traditional Pashmina shawl

and their settings into unique patterns with neutral color palettes in the background to make them more attractive. Computer-aided design software programs have been used to manipulate and intensify these motifs to use as an archival data for such manufacturing. However, the technique is the old one so the essence could be kept intact [16].

Fig. 13 Traditional Indian woolen Kashmiri shawl with nature inspired yet stylized motifs

2.3 *Indigo*

Color dying in India has a long tradition dating back to the time when agriculture was introduced to this land in around 7000–2000 BC [17]. The excavations in India indicate that there was the usage of dying on cotton by the second millennium. Indigo is one of the oldest dyestuffs and the only colorfast organic dying medium which is organic in its nature. The etymology of the word Indigo refers to its origin. It means 'a substance from India' extracted from the Greek word 'indikum', hinting its roots attached to the classical period as well. Writers of that time have indicated its use in cosmetics, medicine and forms of coloring [18]. It was in the seventeenth century when Indian trade started with England and Dutch East India that it got popular due to increase in its demand in the global market [19]. These materials became popular due to its classification as organic dyes originated from vegetable, animal source and mineral dyes [20]. Nowadays, its durability and environmental friendly characteristics allow the designers to use it for surface ornamentation Fig. 14 in various garments and interior accessories.

2.4 *Popular Attires Made of Indian Textile Crafts*

Textile products and garments of India had not only established itself as a craft that markets the cultural identity of the different areas mentioned above but also gives a whole new dimension to ethnic apparel through merging with the regional flavors yet a modern outlook. It is the bread and butter of the local craftsmen and the source of their artistic output and at the same time an industrial domain. It is the one that completes the outlook with any traditional costume worn in different parts of this region. It is a part and parcel of many occasions celebrated across the country.

Fig. 14 The Indigo technique is mostly used on tye dye fabrics which are used in different experimental products

Though its popularity has declined over the years, yet it is still a part of many regions' cultural identity. With its evolving pattern and journey from the traditional legacy to its cultural identification, one can see the changes and the diversification that makes it so intriguing.

Some of the more popular costumes commonly worn in this region which are ensemble with these textile produce are the following.

2.4.1 Suthan and Pehran

This is a popular traditional dress, mostly worn in urbanized Hindu culture on formal or special occasions. However, on the countryside, it is worn on daily basis as an everyday outfit.

Suthan is a form of trousers worn by both male and female. It is baggy and heavy at its bottom part and has mostly embroidered cuffs (paichas). Its form is inspired by a Patiala shalwar and draping technique is like the dhoti of Hindus. A belt, drawstring or a Kamarband is used to hold the shalwar at the waist. The lower class wear an indigo-colored Suthan mostly made of cotton. While the higher class wear a white or black Suthan made with silk or satin.

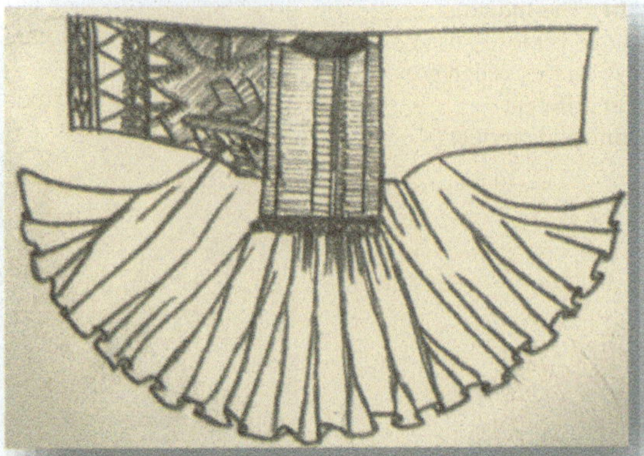

Fig. 15 Pehran is a famous traditional attire used for a variety of local embroidery work

The Pehran (Fig. 15) or the shirt comes with a variety of embroideries and colors in all types of sizes. Initially, the shirt used to be loose and without cuffs but as its popularity increased it became famous on formal occasions as well, carried with baggy sleeves and cuffs. The material mostly used for it is thin muslin and cotton. It had an opening in front which was kept on the right for the Muslims and the left for the Hindus.

2.4.2 Peshwaz, Kurta, and Paijama

Peshwaz (Fig. 16) was a seventeenth century dress, which became popular amongst the women of the royal Mughal court. It was more like a Jama, a full robe dress for a woman. It was usually made of colored muslin and decorated with gold or silver trimmings or borders. The upper half of the dress bodice was also ornamented richly. The Peshwaz had long tightly fitted sleeves. These Peshwaz are worn with printed Jamawar Paijamas (Fig. 17) and long Udhanis. The Kurta is straighter and the patterns are stitched together to form a tunic. There is rich intracultural significance of Peshwaz where strong influences of other dynasties like the Rajput Mughal and Nizam cultures could be detected in its evolutionary stages [21].

The Paijama another unisex garment is popular in a variety of materials like satin, cotton, or fine silk. Another variety of Paijama is a Churidar Paijama popular in this region. The straight-cut pattern trouser is known as Paijama while the one with the tight fitting is called the Churidar Paijama. This type of attire is preferred for female brides nowadays or is worn on formal occasions, so the khussa, worn with it is mostly embellished with Zari or Kundan.

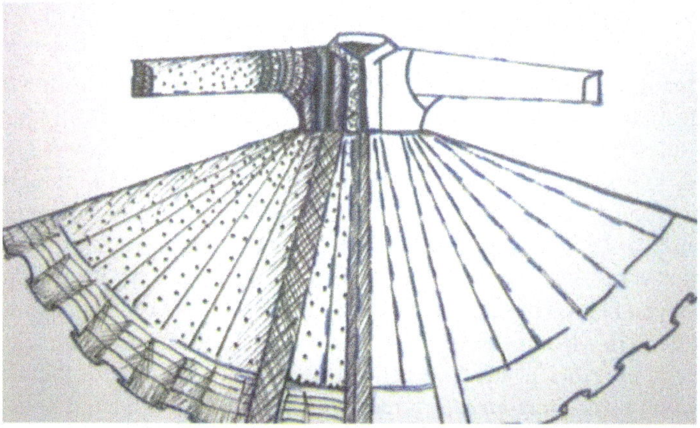

Fig. 16 Peshwaz, was a popular attire worn on festive seasons and the royal court in the past

Fig. 17 Paijama is mostly worn with a Kurta or a Peshwaz

2.4.3 Churi, Shalwar, and Udhani

There are many types of groups living in this region of Tharparker. The low-lying delta areas of the Indus Larh are the native regions to Jat nomads who live in small communities and survive through the breeding of camels, buffaloes, and cows. These groups have certain similarities in their dressing, yet they have the basic cultural differences to showcase their identity [22]. These groups are Malkani, Nuhani, Jumani, Farqani, and Umrani Jats. The basic dressing of the women in the group is a fully rounded skirt whose length is below the knees. The pleats or the gathers of the Churi start from the waistline of just behind the armholes. Beneath the neckline is a Gaj, a piece of ornamentation done in a symmetrical pattern on the bodice as a decorative element. The silhouette and the color of the Churi show the status of the tribe of the women. Red Churis are worn by young girls or brides with colorful embroidered khussas. While the widow wears Churis in relatively plain colors. Gathered fully rounded trousers are worn with the Churis. A large patterned Udhani is also a part and parcel of this attire. The women other than the Jut's caste wear a slack tunic which is extended below knees. It is complemented with a Susi stripped pattern Shalwar [23].

The colors that are worn are mostly of red, green, indigo, orange, purple, white and black known as a Satrangi palette.

2.4.4 Choli, Ghagra, and Udhani

These four elements Choli, Ghagra (Fig. 18) and Udhani was borrowed tradition initially from the harams of the nawabs and the royal courts and then the Rajasthani culture. It is typically worn in the dull background of the deserts with a colorful striking palette. Ghagra is a lengthened skirt usually embroidered with regional Sindhi patterns. An essential feature of the Ghagra is the drawstring or the Izarband which hangs on the sides of the Ghagra decorated with pompoms or Jhumkas. The flare and the volume of the Ghagra give a splendid sight when the women dance and reveal their tradition footwear. Choli and Udhani are also a part of this assemblage. At times the Choli is also replaced by a small kurti. It is adorned with heavy embroidery or silver or gold work on formal occasions.

2.5 Generations: Ages and Conceptual Connotations

Design's role has changed through generations from product to performance. Thus, design in this sense goes beyond the traditional practice of the object or the concept. The fixed identity associated with a design object has become dismantled to a design process; thus, identifying the consumer here is also not a fixed preposition rather a pathway [24]. With such a rich heritage of textiles and crafts in the Indian region and with so many manifestations it is difficult to position it in the present world as

Fig. 18 Ghagra is a popular attire of this region with a history in the Nawab and the Rajasthani culture

only a textile produce or a craft. If design of a textile based craft cannot be linked to a fixed subject anymore or it is not identified as a product, or a tool then where can we categorize it?

One of answers to this discussion could be that it is a process of transformation, with different ups and downs, jumps, in losing and gaining its identity and character over the years. The space within this transition is to be observed. Whereas, as an art piece it is defined visually, culturally, and aesthetically bringing a lot of dialogue to its existence in the past practices of our craft history. It can take sides within our social inhabitant, telling a lot about the political stature, traditional reformation, and our culture. It has also showcased different elements with the ones that can be abandoned or reserved, from what we inherit. There is this point, a fine line in between both entities, for the viewer as an artist to perceive it as a craft or an art piece. In this observation, if these textile heirlooms belonged to any other part of the world like China, England or Europe, its attributes and the process might qualify it to be an art piece, as the viewers were not the ones who had a direct affiliation with it. Due to its repeat value, frequent use, and availability, we have categorized it as a craft only, not cherishing the real meaning of it.

Referring to the above discussion, if we summarize the argument, contemporary design practices now presents the raw ground for an engaging narrative building activity where not only new ideas could be evaluated but also question conventional ways of retrieving our cultural entities. Indian indigenous textiles, here as a product, craft and artifact opens an interesting dialogue to what our societal produce has gained, nurtured and at the same time refrained, dejected, and ignored. The hybridization and the adaptive quality of this craft has left a space where every stakeholder, inspiration and user are gaining from it what they want to perceive. Its origin lies in a design context in which the product design is being built on conditional grounds of the user, culture and the producer.

3 Bangladeshi Craft

Bengali textiles had for years been eminent for its fine quality and vibrant prints. The rich vibrant culture of Bangladesh is a great resource for a wide range of textile crafts produced by legendary artisans who try to reflect an array of literature and folk heritage through their skillful manufacturing techniques. Crafty products are an important source of livelihood and employment after agriculture in Bangladesh. These crafts offer a variety of handwoven textiles made from handspun and hand-dyed fabrics, printed and embellished by hand to fast fashion apparels. Some of the most popular techniques are natural dyes, block printing, handloom textiles, and unique embroideries. The techniques and the byproducts of handwoven textiles could be differentiated by location and different products [25].

3.1 Sustainable Textiles

The manufacturing of textiles in an environmentally and socially viable manner, taking into account factors such as water, energy usage, waste management, and worker's working ethics, is considered sustainability. Bangladesh has been playing a key role in improving such sustainable practices to meet the requirements of the global market. Some of these measures taken by the Bangladeshi industry include the usage of organic cotton yarns, ecofriendly dying processes, and labor laws. They are also exploring ways to reduce their carbon footprint and improve working conditions [26].

3.1.1 Natural Dyes

In the past, dying techniques were associated with different themes extracted from individual, ethnic, and religious connotations. The expression of the craftsman was a conscious effort to reflect himself onto the products. These hand-crafted woven cloth was mostly used on special traditional occasions. However, in the nineteenth century with the introduction of synthetic dyes, this cottage industry nearly dimensioned with the other options gaining more popularity [27]. It was in the late twentieth century when organic dying started to retrieve with researchers like Ruby Ghuznavi taking interest in possibilities to simplify these processes. The aim was to combine contemporary printing techniques with traditional dying methods by simplifying the processes and giving more recognition to the local craftsman.

Dyes are mostly made from natural resources like black color is extracted from the boiling of the bark of the Kala Gab tree and red is prepared from the root of Rong Gachh tree. Blue Indigo color is achieved by the soaking of the indigo leaves for at least 2 days in an earthen vessel. Another color achieved by the mixing of the turmeric powder and the mango tree bark is the yellow color. Although these yarns are purchased from the market but they need to be given a certain treatment before the process of weaving [28].

3.1.2 Block Printed Textile Products

Block printing is a 355-year-old technique. In this technique, wooden blocks are chipped off to replicate a design of a motif or an ornament. Further, then these designs are stamped onto the fabric to recreate beautiful patterns (Fig. 19) with repeating it in a certain compositional method. In Bangladesh, these patterns are recreated on many different products ranging from Shalwar Kameez (traditional shirt and a trouser), men's collared shirts to kids wear. They are also used on bed linens and table cloths for household usage.

These designers also experiment with discharge block printing as it is often designed on apparels where the complete fabric is immersed in color, including the particular spaces on which design is to be formed. Next it is countered with a

Fig. 19 Background of wax printed surface, Batik work

Fig. 20 Woven fabrics had been experimented with a variety of techniques in Bangladesh to achieve a certain texture (Left) or as a base cloth (Right)

chemical compound to make it bleached from that particular part and then the faded portion is colored with a different dye.

Mostly Muslin fabric is used for hand block printing and usually adorned with floral organic motifs. Designers and brands have a wide variety of such motifs and patterns for different articles. They also combine hand loom and block printed techniques together to create a variety of designed block printed hand bags, backpacks, tableware sets, and furniture items.

Multiple boutiques and silk houses have collaborated with designers to come with block printed silk saris where the primary material is the Balaka silk. The products other than the saris are Shalwar Kameez, Fatuas, and Kurtas. As this trend has extended other luxury, casual and semi casual products have added to this list catering both the male and female target group.

3.1.3 Wax-Printed Fabrics

Another method similar to block printing technique is wax printing (Fig. 20) where wax is used instead of ink on the block. The wax repels the dye to create the print onto the undyed surface of the fabric. This practice is usually done on dark-shaded fabrics with lighter dyed imprints on it. This fabric is then used for multiple products like bed linens and apparel.

3.1.4 Cotton and Silk Handloom Textile Products

Before the colonial rule in Bangladesh, fine muslin was made in Dhaka and exported all over in Asia and Europe. The quality of the muslin was so fine that 11 yard of this fabric could be wadded in one hand. According to another folk tradition, sari could be drawn from a small ring due to its finesse in weaving [29]. According to the local

Fig. 21 Woven fabrics had been experimented with a variety of techniques in Bangladesh to achieve a certain texture (left) or as a base cloth (right)

craftsmen, it was the favorite cloth of the aristocracy and the royals. A transparent Mulmul cloth named Gangetic muslin was used in the subcontinent. It was very lightweight and referred to as nebula, mist, or vapor in the Greek or Roman writings. In Dhaka, it is still named as Venti or woven fabrics due to its popular trend [30]. Woven textiles (Fig. 21) are used in a variety of textile products to create textures or fine surfaces.

3.2 *Traditional Attires of Bangladesh*

These traditional attires representing the cultural heritage of Bangladesh have an old legacy. It is the representation that could be seen on occasions like Pohela Boishakh (Bengali New Year), Durga Puja, Eid-ul-Fitr, and wedding scenarios. It is interesting to observe the cultural background of Bangladeshi festivals that has five important features that highlights their contribution in sustaining their textile crafts. These factors are devotedness, tolerance, simplicity, aestheticism, and cultural restoration. They are reflective in their attires as well [31].

3.2.1 Sari, Lungi, and Kurtas

Sari is the national dress of Bangladesh, and Dhakai Jamdani is one of the famous types in this collection. They usually pronounce them shares and one can find a huge variety of it in different materials like the georgette and silk too [32]. Banarasi and Katan are also popular types of saris in the Bangladeshi region. Historically, the artisans who weave these saris belonged to the region of Banaras in India. These weavers are natives of Bihar who migrated to East India during the 1947 movement.

The data of Bangladesh Handloom Board provide information about these immigrants who established the Banarasi industry, Mirpur in around 1950. These weavers created a variety of Jamdani, Mala, and Qatan sarees and accessories especially for semi-formal to formal occasions [33].

Lungi is a common man's garment for mostly informal wear in the Bengali culture. In the past Rohitpur, a region 20 km near Dhaka, it was famous for the making of this product but now there is rarely any cottage industry there. Now it has moved to Enayerpur and Serajganj district with a higher quality manufacturing cottage industry. The Lungi is mostly available in multicolor with a very unique pallet [25]. It is traditionally made with checks and boxes with intersecting lines as patterns, mostly printed. It is usually manufactured with 100% cotton and sometimes with mixed yarns. Bengali Lungi business has garnered a special position in the international market especially in India and Indonesia due to its fine quality and bright colors. It is quite famous on festive occasions and, in the past, it was considered a sign for nobility.

3.3 Design Narratives with Sustainable Goals

Over the past years and the present set of design discoveries, with wonderful modes of collaborations with other art and design forms, it has been a willingly blind nod to the freedom of creativity and potential in myriad forms of these textile crafts by its consumers. Globalization, consumerism, mass media, advertisement, pluralism, and hybridity all played a key role in either evolving it or at the same time declining its original landmark quality as a craft form. Because of this progressive time in terms of industry and technology people were also offered with other variety of substitute synthetic textiles in the market. However it bought with it, its share of extreme water pollution, emissions, and increased carbon footprint, in spite of using several recycling techniques [34]. But with the growing numbers of environmental challenges faced by the societies and the nations, it became inevitable to avoid certain suitable environmental friendly practices. These textile crafts with the use of such vegetable dyes, hand manufacturing techniques and less waste are ideal solutions to such a problem. Although many fashion brands and designers have taken this task as challenge and created their own design narratives with sustainable goals keeping in loop theses traditional textile practices [35] still a lot could be done to meet the international requirements.

4 Conclusion

Looking at the journey of textile products as a communication medium of the value sets of the past heritage, we can observe it with queries like, what people wear speaks to what they hold important. What does this craft wants to translate in terms of heritage? How it communicate to the world in terms of hybridized products? It was observed that they represents value, craft [35], and longevity, but also subjective mystical values, like aesthetic and status. Such perspective could carry a different meaning to the wearer, and another to the world, but generally there is an unspoken, shared understanding. The extensive, highly ornamented, rich textile craft legacy sends a socially shared message of indulgence and luxury that was enjoyed by the ancestors of the past. It also communicates the more abstract concept of ethnicity to the textile sector of the past in our region, which was built over time through social and media constructions of identity. Sense of power, regality and a set of connotations associated with it, with a large gap of recorded authenticated history that could have been a greater resource to ascertain its value as a craft. But designers, small enterprises and a large cottage industry retail network played along with these established codes and tried to enrich the tale of the human expression rooted to this region. These codes hinting to the past culture in the form of ornamentation, manufacturing techniques and sustainable materials. They had been direct and clear to engage a larger range of consumers into buying it. But it is interesting how this role play is creating a diversification through introducing it as a medium or an expression by small local enterprises or social media platforms. They are using it as an obvious form of fashion communication, in contrast to more commercial forms of it available in market. The expression becomes so subtle that the mass audience might not react to it, but the object expresses values, though less use of value and more mystique and aesthetics involved to the point that many consider it as a form of art. This is the realm of appreciation for such an object like textile crafts where the glamour and even the legacy is negotiated by these designers or brands to define this medium in front of the larger audience locally and internationally.

However, on the whole, it is interesting to see how these diversities in the textile products of these regions have added this sector as an important beneficiary in their economic growth as well. As India, Pakistan, and Bangladesh share the same geographical and historical legacy thus, a lot of similarities are there in terms of cultural exchange and artisanal skillset. Yet industrial interventions and modern adaptions have changed these traditional textile forms into variations catering to the needs of the modern target market. As it is more time consuming and have less economic benefits so the younger generation of artisans are not interested to acquire or practice these skills. Also, apart from low wages, shortage of raw material or its lack of proper development is another reason for its downfall. The need of the hour is to preserve these textile crafts parallel to the development of textile industry in this region. The need is to hybridize them into meaningful crafts and textile products.

References

1. Leslie CA (2007) Needlework through history: an encyclopedia. USA, Greenwood Publishing Group
2. Barber EJ (1994) Womens work: the first 20000 years women cloth and society in early times. WW Norton, New York
3. Carr T (2014) The harappan civilization by Tarini Carr. Archaeology online. Web
4. Nadiem IH (2002) Portrait of Sindh. Sang-E-Meel Publications, Lahore
5. Dani AH, Masson VM (2003) History of civilizations of central Asia: development in contrast: from the sixteenth to the mid-nineteenth Century. UNESCO, Paris
6. Seiko Jose ST (2022) Handbook of museum textiles, volume 1: conservation and cultural research. Wiley, USA
7. Taknet DK (2016) Jaipur: Gem of India. Integral DMS, India
8. Naik SD (1996) Traditional embroideries of India. APH Publishing, India
9. Gillow J, Barnard N (2008) Indian textiles. Thames & Hudson, UK
10. Wynbrandt J (2009) A brief history of Pakistan. Infobase Publishing, USA
11. Hughes AW (1874) A Gazetteer of the Province of Sindh. G. Bell & Sons, London
12. Asher FM (2003) Arts of India, from prehistory to present. Britannica Encyclopedia, Inc., Hong Kong
13. Zeiler XY (2018) In: Jain P, Sherma R, Khanna M (eds) Hinduism and tribal religions. Encyclopedia of Indian religions. Springer, Dordrecht
14. Weibel AC (1942) Indian textiles. Bull Detroit Inst Arts City Detroit 21:52–53
15. Chatterjee R (1912) The modern review, vol 12. Modern Review Office, Calcutta
16. Guru R, Thennarasu P, Kumar R (2022) Study on the traditional handloom textiles in India. TextE Leather Rev 5:392–413
17. Punia R (2018) Dyeing in ancient Indian textile : an analytical study. Ascent Int J Res Anal 3:28.1–28.6
18. Balfour-Paul J (1999) Indigo in South and South-East Asia. Text Hist 30:98–112
19. Nadri GA (2020) Indigo in precolonial South Asia. In: Oxford research encyclopedia of Asian history. Oxford University Press
20. Kamat SY, Alat DV (1990) Natural dyes: a dying craft? Indian Text J 100:66–70
21. Ayachit, S. (2023). Aambkapad, Chira, Khun, and more–intercultural influences on fabrics of the Peshwas of Pune, India. Text: J Cloth Cult 21:340–351
22. Ewers JC (1980) Climate, acculturation, and costume a history of women's clothing among the Indians of the Southern Plains. Plains Anthropol 25(87):63–82
23. Singh OP, Sidhu R (1985) Costumes in India
24. Wani NS (2022) Factors influencing price perception for fashion: study of millennials in India. Vision 26:300–313
25. Parishada BJ (2003) Textile traditions of Bangladesh. National Crafts Council of Bangladesh, Minnesota, p 7
26. Hossain MI, Ong TS, Siow ML (2022) Nexus of stakeholder integration, green investment, green technology adoption and environmental sustainability practices: evidence from Bangladesh textile SMEs. Pertanika J Soc Sci Humlties 30:253–281
27. Islam MM, Khan A, Islam M (2013) Textile industries in Bangladesh and challenges of growth. Res J Eng Sci 2:33–42
28. Mahbub T, Minhus SM, Hui T (2021) Fashion of Indigenous people in Bangladesh: exploring traditional clothing process of Chakma Tribe. J Textile Sci Fashion Tech 9(1)
29. Nevins D, Whyte M (2018) Bangladesh. United States: Cavendish square 106
30. Parker PM (2011) The 2009 import and export market for men's and boys' suits of woven textile fabrics in Bangladesh. Country trade reports
31. Minhus SM, Huie L (2021) The tendency of traditional costume at heritage festival for cultural revival. SAGE Open 11
32. Urmi Rahman CS (2014) Bangladesh—culture smart! the essential guide to customs & culture. Kuperd, Great Britain

33. Ferdous J (2014) Mirpur Benarasi Palli of Bangladesh: living and livlihood. Int J Chem Process 87–97
34. Hossain MI, Ong TS, Teh BH (2024) The panorama of corporate environmental sustainability and green values: evidence of Bangladesh. Environ Dev Sustain 26:1033–1059
35. Asif AKMAH (2017) An overview of sustainability on apparel manufacturing industry in Bangladesh. Sci J Energy Eng 5:1

Micro and Macroeconomic Overview of Creative and Value-Added Textiles

Asif Mahbub Karim and Allah Dad

Abstract Textiles, often considered the fabric of civilization, have transcended their utilitarian origins to become a canvas for creative expression and cultural representation. In the realm of design and craftsmanship, the interplay of innovation and tradition has given rise to textiles that not only serve functional purposes but also tell compelling stories.

1 Introduction

Textiles, often considered the fabric of civilization, have transcended their utilitarian origins to become a canvas for creative expression and cultural representation. In the realm of design and craftsmanship, the interplay of innovation and tradition has given rise to textiles that not only serve functional purposes but also tell compelling stories. This write-up explores the fascinating world of creative and valued textiles, examining how artisans and designers weave together threads of tradition with contemporary ingenuity to create unique and cherished pieces.

As we move into the contemporary era, textile artists and designers are pushing the boundaries of tradition through innovation in both materials and techniques. The exploration of unconventional fibres, such as recycled materials, bamboo, and even conductive textiles, reflects a commitment to sustainability and a departure from conventional norms. In the realm of technique, digital technology has revolutionized textile design. Computer-aided design (CAD) and digital printing allow for intricate and precise patterns that were once unimaginable. This fusion of technology with traditional craftsmanship is evident in the work of contemporary textile artists who seamlessly blend the old and the new.

A. M. Karim (✉)
Dean, Binary Graduate School, Binary University of Management & Entrepreneurship No.1 IOI Business Park, Persiaran Puchong Jaya Selatan, Bandar Puchong Jaya, 47100 Puchong, Selangor, Darul Ehsan, Malaysia
e-mail: drasif@binary.edu.my

A. Dad
Pakistan Institute of Fashion and Design, Johar Town, Lahore, Pakistan

H. Jamshaid and A. Dad (eds.), *Creative Textile Industry*, SDGs and Textiles,
https://doi.org/10.1007/978-981-97-3802-1_3

Textiles have become a medium for artistic expression, blurring the lines between fashion and art. Renowned artists collaborate with fashion designers to create wearable art, turning garments into canvases that embody both aesthetic and conceptual depth. The avant-garde creations of designers like Alexander McQueen and Iris van Herpen exemplify the transformative power of textiles.

The value of textiles extends beyond their aesthetic appeal. Handcrafted textiles often carry a story of labour, skill, and cultural significance. Artisanal communities around the world produce textiles that are not only visually stunning but also imbued with a sense of authenticity and heritage.

Fairtrade practices and ethical sourcing have become integral to the valuation of textiles. Consumers are increasingly conscious of the social and environmental impact of their purchases, leading to a growing demand for textiles that are ethically produced and sustainably sourced.

Examples of creative and valued textiles can be seen in, for example, firstly, Kantha Embroidery of Bengal, India. Kantha, a traditional form of embroidery from Bengal, involves stitching together layers of old sarees to create quilts and garments. Each piece tells a unique story through its hand-stitched patterns, reflecting the history and emotions of the artisan. Secondly, Aran Knit Sweaters of Ireland. Aran sweaters, with their intricate cable patterns, originated from the Aran Islands in Ireland. Each stitch is symbolic, representing elements of island life and serving as a form of storytelling through textiles.

Thirdly, we can see the Japanese Boro Textiles. Boro, a Japanese term meaning "rags," refers to textiles that have been mended and patched over time. Originally born out of necessity, Boro textiles have evolved into a form of art, embodying the Japanese philosophy of wabi-sabi, celebrating imperfections and the passage of time.

Innovations in textile technology have given rise to smart textiles that integrate functionality with fashion. Fabrics with embedded sensors, LED lights, and even temperature-regulating properties showcase the potential for textiles to go beyond aesthetics and serve practical purposes in everyday life.

Creative and valued textiles are a testament to the ever-evolving nature of human creativity and craftsmanship. The intersection of tradition and innovation in the textile world not only produces visually stunning pieces but also preserves cultural heritage and promotes sustainable practices. As we continue to weave the threads of the past with the possibilities of the future, textiles will remain a dynamic and cherished form of artistic expression, connecting us to our roots while propelling us forward into uncharted territories of design and craftsmanship.

2 Background

Textiles have been an integral part of human history, serving as more than just garments. From the intricately woven linens of ancient Egypt to the vibrant silk textiles of Imperial China, each culture has left its indelible mark on the world of textiles. Traditional textile techniques, passed down through generations, embody cultural identity and heritage. The history of textiles is a colourful and intricate tapestry, woven with threads of tradition, innovation, and artistic expression. From ancient civilizations to contemporary societies, textiles have played a pivotal role in reflecting cultural identity, craftsmanship, and the evolving tastes of humanity.

3 Ancient Foundations: From Flax to Silk

The origins of textiles can be traced back to prehistoric times when early humans utilized natural fibres like flax and animal hides to fashion rudimentary garments. As civilizations flourished, so did the sophistication of textile production. Ancient Egypt, renowned for its mastery of linen, produced exquisite fabrics that adorned both the living and the deceased. The Silk Road, connecting East and West, facilitated the exchange of silk from China, revolutionizing the luxury textile market and sparking a global fascination with this sumptuous material [73].

4 Cultural Textile Heritage: A Global Kaleidoscope

As civilizations evolved, each culture contributed its unique flavour to the world of textiles. Weaving, a fundamental human craft, boasts a rich history stretching back over 12,000 years to the Neolithic era. Archaeological evidence, such as fragments of woven cloth and looms, underscores its early existence and significance in ancient societies. For instance, discoveries in sites like Çatalhöyük in Turkey and the Swiss lake dwellings provide tangible proof of early weaving practices [6].

The advent of agriculture during the Neolithic period precipitated the development of weaving, as settled communities cultivated fibres from plants like flax and cotton, and domesticated animals for their wool. These fibres were spun into threads and then woven into textiles for various purposes, including clothing, shelter, and trade [53]. Moreover, weaving played a crucial role in the cultural and economic exchange among ancient civilizations. The intricate patterns and techniques developed by different cultures, such as the warp and weft methods of Mesopotamia or the intricate tapestries of the Andean peoples, illustrate the diversity and sophistication of early weaving traditions [41].

The intricate embroidery of China, the indigo-dyed fabrics of West Africa, and the vibrant ikat weavings of Central Asia all stand as testament to the diverse techniques and aesthetics that have been passed down through generations [12].

India, with its millennia-old tradition of handloom weaving, showcases a rich tapestry of textile heritage. From the intricate Banarasi silk sarees to the tie-dye artistry of Bandhani, Indian textiles are imbued with cultural symbolism and exquisite craftsmanship. The Gharchola of Gujarat, worn during auspicious occasions, is an excellent example of how textiles become intertwined with rituals and ceremonies, reflecting the values and beliefs of a society [32].

5 Innovation in Textile Techniques: From Looms to Jacquard

The medieval period witnessed significant advancements in textile production techniques. The use of vertical looms, treadle looms, and the spinning wheel revolutionized the efficiency and output of textile workshops. Basile Bouchon, Jean-Baptiste Falcon, and Jacques Vaucanson each made significant contributions to the development of programmable looms, laying the groundwork for modern automation in textile manufacturing. Basile Bouchon, a French inventor, is credited with creating the first known example of a programmable loom in 1725. Bouchon's design utilized a perforated paper strip to control the weaving pattern, allowing for greater precision and efficiency in textile production [21].

Jean-Baptiste Falcon, a fellow French inventor, further advanced the concept of the programmable loom in the 1720s. Falcon improved upon Bouchon's design by introducing punched cards made of sturdy materials like cardboard or metal, which could be easily manipulated to alter the weaving pattern [22].

Jacques Vaucanson, another French engineer and inventor, made significant strides in automating textile manufacturing during the eighteenth century. Vaucanson's most notable contribution was his development of the first fully automated loom in 1745, known as the "Jacquard loom." This innovative machine used a series of punched cards to control the weaving process, allowing for complex patterns to be produced with ease [74].

Collectively, the contributions of Bouchon, Falcon, and Vaucanson revolutionized the textile industry, paving the way for modern automation and laying the foundation for subsequent advancements in manufacturing technology. The Industrial Revolution further transformed the textile industry, bringing mechanization and mass production. Cotton mills in England and textile factories in New England became the epicentres of a booming industry, albeit with significant social and economic consequences [43].

6 Revival of Handcrafted Traditions: Artisanal Resurgence

Despite the industrialization of textiles, the late nineteenth and early twentieth centuries witnessed a renewed appreciation for handcrafted traditions. The Arts and Crafts Movement, led by figures like William Morris, championed the revival of traditional craftsmanship and rejected the dehumanizing effects of mass production [55].

This movement laid the groundwork for the preservation and revitalization of artisanal textile techniques. Textile designers like William Morris, Mariano Fortuny, and the Wiener Werkstätte embraced the beauty of handmade textiles, emphasizing the uniqueness of each piece [19].

7 Contemporary Intersection: Art, Fashion, and Technology

In the contemporary era, textiles have transcended their utilitarian origins to become a medium for artistic expression and technological innovation. The convergence of art and fashion has given rise to wearable art, where designers collaborate with artists to create garments that transcend the boundaries of traditional fashion [26]. Advancements in technology, particularly in digital design and fabrication, have opened new avenues for textile innovation. 3D printing, smart textiles with embedded sensors, and sustainable materials are shaping the future of textiles, marrying aesthetics with functionality [58].

8 Valued Textiles and Ethical Consumption

As the global awareness of ethical consumption grows, so does the demand for textiles with a story—pieces that are not just visually appealing but also socially and environmentally responsible. The Fair-Trade movement has gained momentum, ensuring that artisans receive fair wages for their work, and sustainable practices are employed throughout the production process [38]. Consumers are increasingly seeking transparency in the textile supply chain, prompting designers and manufacturers to adopt eco-friendly materials and ethical production methods. This shift towards conscious consumption underscores the value placed on textiles that align with personal and societal values [60].

Creative and valued textiles are the result of a rich historical tapestry interwoven with cultural heritage, innovation, and ethical considerations. From the ancient looms of Mesopotamia to the modern studios of sustainable fashion designers, textiles continue to evolve, reflecting the changing needs and values of society. The appreciation for traditional craftsmanship, coupled with a commitment to ethical and sustainable practices, ensures that the world of textiles remains a dynamic and cherished realm, where the past and the present are intricately connected in a seamless weave of creativity and cultural expression.

A. **Macroeconomic Overview**

The creative and valued textiles industry plays a crucial role in the global economy, encompassing a diverse range of products that blend artistry, tradition, and innovation. This macroeconomic overview explores the significance of the creative textiles sector, examining its historical evolution, current economic landscape, and future prospects. From traditional handcrafts to cutting-edge innovations, textiles contribute not only to fashion and art but also to employment, trade, and sustainable development.

9 Historical Context: From Handlooms to Global Trade

The history of textiles is intertwined with the economic development of nations. Traditionally, textiles were produced through labour-intensive processes, often within local communities. With the advent of the Industrial Revolution, mechanization transformed the industry, leading to mass production and the establishment of global trade networks [48]. The economic impact of textiles during the Industrial Revolution was profound, driving economic growth and urbanization. Textile mills became key drivers of industrialization in regions like England and the Northeastern United States. The economic importance of textiles extended beyond manufacturing, influencing trade patterns, labour markets, and even international relations [44].

10 Current Economic Landscape: Global Markets and Supply Chains

In the contemporary era, the creative and valued textiles industry is a multifaceted global enterprise. The value chain is extensive, involving raw material production, manufacturing, design, distribution, and retail. Countries around the world contribute to different stages of this chain, creating a complex web of global interdependence [37].

Manufacturing Hubs

Developing countries, particularly in Asia, have emerged as major manufacturing hubs for textiles and apparel. Nations like China, India, Bangladesh, and Vietnam have significant textile industries, benefitting from lower production costs and skilled labour. These countries play a vital role in meeting global demand for a wide range of textiles.

Design and Innovation

Creative textiles are not only about production but are also about design and innovation. Countries like Italy, France, and Japan are renowned for their contributions to high-end fashion and textile design. Innovation in textile technology, sustainable practices, and artistic expression are key drivers of economic value in these regions.

Trade Dynamics

The global textile trade is intricate, involving the exchange of raw materials, intermediate goods, and finished products. International trade agreements, such as the World Trade Organization (WTO) agreements, impact the textile industry by influencing tariffs, quotas, and trade regulations [72].

11 Employment and Economic Development: Textiles as Job Creators

The creative textile industry has historically been a significant source of employment, particularly in developing economies. From the skilled artisans producing hand-woven textiles to the factory workers involved in mass production, textiles contribute to job creation at various skill levels.

Artisanal Employment

Handcrafted textiles, often associated with cultural traditions, provide livelihoods for numerous artisans worldwide. Whether it's weaving, embroidery, or traditional dyeing techniques, these crafts support local economies and preserve cultural heritage [59].

Mass Production Jobs

The rise of industrialization brought about mass production, leading to the establishment of textile factories and assembly lines. While this created jobs on a large scale, it also raised issues related to labour conditions, workers' rights, and fair wages [25].

Global Supply Chains and Labour Mobility

The global nature of the textile industry has implications for labour markets. Workers in manufacturing hubs often migrate from rural areas to urban centres in search of employment opportunities. This movement contributes to urbanization but poses challenges related to living conditions and workers' rights.

12 Sustainability and Ethical Consumption: A Growing Market Trend

In recent years, there has been a paradigm shift in consumer preferences towards sustainable and ethically produced textiles. This shift is not only driven by environmental concerns but also by a desire for transparency and social responsibility in the production process [30].

Eco-Friendly Practices

Sustainable and eco-friendly textiles are gaining traction, with a focus on reducing environmental impact. This includes the use of organic and recycled materials, eco-friendly dyeing processes, and adherence to circular economy principles.

Ethical Sourcing

Ethical considerations in the supply chain, such as fair wages, safe working conditions, and the absence of child labour, have become crucial factors in consumers' purchasing decisions. Certification programs, like Fair Trade, play a role in promoting ethical sourcing.

Consumer Awareness

Increased awareness among consumers regarding the social and environmental impact of their purchases has led to a demand for transparency. Textile companies that align with sustainable and ethical practices often find themselves in a favourable position in the market [4].

13 Challenges and Opportunities: Navigating a Dynamic Landscape

The rapid industrialization of the textile sector has historically brought about a myriad of negative social consequences, impacting communities, workers, and the environment. As one of the earliest industries to undergo significant mechanization, the textile sector offers a rich case study of the social ramifications of rapid industrialization.

Firstly, in the context of labour, rapid industrialization in the textile sector often led to exploitative working conditions and low wages. During the early phases of industrialization, textile factories employed women and children in large numbers, subjecting them to long hours, dangerous working conditions, and meagre pay. The infamous textile mills of the Industrial Revolution in Britain serve as a poignant example, where workers, including children as young as five, toiled in hazardous environments for minimal compensation [1]. This exploitation of labour contributed to social unrest and the rise of labour movements advocating for workers' rights.

Moreover, rapid industrialization in the textile sector has frequently led to environmental degradation. Textile manufacturing processes involve the use of chemicals, water, and energy, contributing to pollution and resource depletion. From the discharge of toxic dyes and chemicals into water bodies to the emission of greenhouse gases from energy-intensive production processes, the environmental footprint of the textile industry has been substantial [27]. This pollution has not only harmed ecosystems but also affected the health and livelihoods of communities living in proximity to textile factories.

Furthermore, the rapid expansion of the textile industry often resulted in the displacement of traditional livelihoods and communities. In regions where textile production became dominant, traditional artisanal practices were marginalized or replaced by mechanized production methods. For example, the rise of textile factories in India during the colonial period led to the decline of indigenous handloom weaving, disrupting centuries-old cultural and economic practices [16]. This displacement contributed to social upheaval and the erosion of cultural identities.

Additionally, the socio-economic impacts of rapid industrialization in the textile sector have been unevenly distributed, exacerbating inequalities within and between societies. While industrialists and entrepreneurs profited from the expansion of the textile industry, workers often faced precarious employment, low wages, and inadequate living conditions. This economic disparity fuelled social tensions and widened the gap between the wealthy elite and the working class, leading to social unrest and conflict. While the creative and valued textile industry presents numerous opportunities, it also faces several challenges that require strategic considerations.

Fast Fashion and Overconsumption

The rise of fast fashion, characterized by rapid production cycles and disposable clothing, poses environmental challenges and contributes to overconsumption. Balancing economic growth with sustainable practices remains a delicate task [28].

Technological Disruption

The integration of technology, while driving innovation, also poses challenges for traditional manufacturing. Automation and artificial intelligence have the potential to reshape the industry, impacting both employment and production processes [10].

Global Economic Shifts

Economic shifts, such as the rise of protectionist policies and geopolitical tensions, can disrupt global supply chains and trade dynamics. Uncertainties related to tariffs and trade agreements can have implications for the textile industry.

The creative and valued textile industry stands at the intersection of art, culture, and commerce, shaping global economies and societies. From its historical roots in traditional craftsmanship to the challenges and opportunities presented by modern economic dynamics, textiles continue to evolve. A nuanced understanding of the macroeconomic landscape is crucial for stakeholders in the industry to navigate the complex web of global markets, sustainable practices, and technological advancements. As the demand for creative and valued textiles grows, the industry's ability to balance innovation with ethical considerations will be essential for long-term success in this dynamic and ever-evolving landscape.

14 Global Key Players in Creative and Value-Added Textile Sector

The creative and value-added textile sector boasts a diverse array of key players, ranging from traditional artisans and small-scale enterprises to globally recognized corporations at the forefront of innovation. These players contribute to the sector's vibrancy, combining craftsmanship, design, and technology to create textiles that are both aesthetically appealing and economically significant. This overview will highlight key players across various aspects of the industry.

Traditional Artisans and Craftsmen

At the heart of the creative textiles sector are traditional artisans and craftsmen who continue to preserve and perpetuate centuries-old techniques. These individuals, often working within local communities, produce handcrafted textiles that are deeply rooted in cultural traditions. From Indian weavers creating intricate Banarasi silk sarees to Japanese craftsmen practising Boro stitching, these artisans are the custodians of heritage and craftsmanship.

High-End Fashion Houses

Internationally renowned fashion houses play a pivotal role in shaping the landscape of creative textiles. Designers and brands such as Chanel, Dior, and Prada are synonymous with luxury and innovation. These houses often collaborate with artists and employ cutting-edge techniques to create haute couture pieces that not only push the boundaries of fashion but also serve as artistic expressions. Their influence trickles down to set trends and standards for the broader fashion industry.

Textile Designers and Studios

Textile designers and studios are instrumental in translating creative visions into tangible fabrics. Names like Liberty London and Marimekko are celebrated for their iconic patterns and prints. These designers often blend traditional craftsmanship with contemporary aesthetics, creating textiles that resonate with modern consumers. The fusion of art and functionality is a hallmark of their work.

Innovative Textile Technology Companies

The intersection of technology and textiles has given rise to innovative companies that are pushing the boundaries of what textiles can achieve. Companies like The Fabricant and Eeonyx specialize in smart textiles, incorporating technologies such as conductive fibres and embedded sensors into fabrics. Their work extends beyond fashion, venturing into areas like health, sports, and even interactive installations.

Global Textile Manufacturers

Large-scale textile manufacturers are pivotal in the mass production of textiles for global markets. Companies like China-based Shandong Ruyi and India's Reliance Industries are major players in the production of a wide range of textiles, from everyday apparel to industrial fabrics. Their operations often span multiple countries, contributing significantly to international trade.

Sustainable and Ethical Brands

With a growing emphasis on sustainability and ethical consumption, several brands have emerged as leaders in this space. Patagonia, for example, is renowned for its commitment to using eco-friendly materials and ethical sourcing practices. These companies prioritize transparency and responsible production, appealing to a consumer base increasingly conscious of the environmental and social impact of their purchases.

Online Platforms for Independent Designers

The rise of e-commerce has provided a platform for independent designers and small-scale enterprises to reach a global audience. Platforms like Etsy and Not Just a Label connect consumers directly with independent designers, allowing for the purchase of unique, handmade textiles. This democratization of access has empowered smaller players in the industry.

Textile Education and Research Institutions

Educational institutions and research centres dedicated to textiles contribute significantly to the sector. Institutions like the Textile Institute and the Fashion Institute of Technology (FIT) foster talent and research in areas such as textile engineering, design, and sustainability. Their work not only educates the next generation of industry professionals but also drives innovation and technological advancements.

Global Retail Giants

Major retail corporations, such as H&M and Zara, have a profound impact on the textiles industry. These fast fashion giants are known for their ability to quickly adapt to trends and deliver affordable, trendy clothing to a mass market. However, they have also faced scrutiny regarding their environmental and labour practices, sparking discussions about the sustainability of fast fashion.

Artists Collaborating with Textile Mediums

Artists who incorporate textiles into their practice contribute a unique perspective to the sector. Figures like Sheila Hicks and El Anatsui use textiles as a medium for artistic expression, creating installations and artworks that transcend traditional boundaries. Their work blurs the line between art and craft, elevating textiles to the realm of fine art.

Creative and value-added textile sector is a dynamic ecosystem with a wide range of key players contributing to its vibrancy. From traditional artisans preserving cultural heritage to global corporations driving technological innovation, each player plays a crucial role in shaping the industry. As the sector continues to evolve, collaboration and dialogue between these diverse players will likely define the future trajectory of creative textiles, balancing tradition with innovation and meeting the demands of an ever-changing global market.

The Industrial Revolution, often associated with the dramatic transformation of economies and societies in Europe and North America during the late eighteenth and nineteenth centuries, was not uniform across all regions of the world. While its impact was profound in some areas, other regions experienced it differently or at different times.

In Europe, particularly in Britain, the Industrial Revolution began in the late eighteenth century, characterized by the mechanization of production processes, the introduction of steam power, and the growth of factory-based industries such as textiles, iron, and coal mining. This period marked a significant shift from agrarian economies to industrialized ones, leading to urbanization, population growth, and technological innovation [48].

North America also experienced significant industrialization during the nineteenth century, with the United States emerging as a powerhouse of industrial production. Factors such as abundant natural resources, a growing population, and technological innovation fuelled industrial growth, leading to the development of sectors such as textiles, steel, and transportation [45].

However, the Industrial Revolution's impact in other regions, such as Asia, Africa, and Latin America, was more varied and often delayed. Colonialism played a significant role in shaping industrialization patterns in these regions, with European powers exploiting colonies for raw materials and markets rather than promoting industrial development [2]. As a result, industrialization in many colonized regions was limited to extractive industries and agricultural production for export, rather than the establishment of diversified industrial economies.

Moreover, cultural and institutional factors also influenced the spread and pace of industrialization across different regions. In Japan, for example, the Meiji Restoration in the late nineteenth century ushered in a period of rapid modernization and industrialization, inspired by Western models but adapted to fit Japanese socio-cultural norms [33]. Similarly, some regions in Latin America experienced industrialization in the late nineteenth and early twentieth centuries, driven by factors such as urbanization, infrastructure development, and state-led industrial policies [11].

B. Market Segmentation in Creative and Value-Added Textiles

Market segmentation in the creative and value-added textile sector is a strategic approach that involves dividing the market into distinct and homogeneous groups based on various criteria. This enables businesses to tailor their products, marketing strategies, and services to meet the specific needs and preferences of different customer segments. In the realm of creative and valued textiles, segmentation is crucial due to the diverse range of products, consumer preferences, and applications within the industry. Market segmentation can be viewed in the following ways.

Firstly, Demographic Segmentation

Age Group

Creative textiles appeal to a wide range of age groups, each with distinct preferences. For instance, younger consumers may be drawn to vibrant, innovative designs and sustainable practices, while older consumers might prefer textiles that reflect traditional craftsmanship and cultural heritage.

Gender

Gender-based segmentation remains relevant in the textile sector. Design preferences, colour choices, and even fabric types may vary between male and female consumers. Some brands specifically target one gender, while others adopt a unisex approach, recognizing the evolving landscape of gender preferences in fashion.

Income Levels

Income is a crucial factor influencing purchasing power and preferences. High-end fashion houses cater to affluent consumers seeking luxury textiles, while budget-friendly brands and independent designers may target a more price-conscious demographic. Understanding income levels helps businesses position their products effectively.

Lifestyle and Hobbies

Lifestyle segmentation considers the varied interests and hobbies of consumers. Outdoor enthusiasts may seek performance textiles with moisture-wicking properties, while interior design enthusiasts might be interested in home textiles that enhance decor. Lifestyle-based segmentation allows for targeted marketing and product development.

15 Secondly, Product Characteristics

Material Type

The type of material used in creative textiles is a significant segmentation factor. For example, some consumers prioritize natural fibres like organic cotton or silk for their comfort and sustainability, while others may prefer high-tech materials such as smart textiles with embedded technology.

Design Aesthetics

Design preferences vary widely, from traditional patterns and cultural motifs to avant-garde and minimalist designs. Segmentation based on design aesthetics allows businesses to tailor their offerings to specific tastes, ensuring they resonate with the intended audience.

Application

Creative textiles find applications across diverse sectors, including fashion, home decor, industrial applications, and art installations. Segmenting the market based on application helps businesses understand the unique needs of each sector and adapt their products accordingly.

Sustainability

With a growing emphasis on sustainable practices, a significant market segment prioritizes eco-friendly textiles. Consumers interested in sustainable fashion and ethical sourcing are likely to favour brands that align with their values, leading to the emergence of a niche market within the larger creative textiles sector.

16 Thirdly, Geographic Segmentation

Regional Preferences

Different regions exhibit distinct preferences in terms of colours, patterns, and cultural symbols. Understanding these regional nuances is crucial for businesses looking to cater to specific markets. For example, textiles with traditional Indian patterns may be more popular in South Asia than in Western markets.

Climate and Seasonal Variations

Climate plays a significant role in textile preferences. Lightweight and breathable fabrics may be more popular in warmer climates, while cold-weather regions may demand textiles with insulating properties. Seasonal variations impact both the design and functionality of textiles.

17 Fourthly, Consumer Behaviour and Psychographic Segmentation

Brand Loyalty and Values

Consumer behaviour often revolves around brand loyalty and shared values. Some consumers may prioritize brands known for their commitment to ethical sourcing and sustainability, while others may be more influenced by cutting-edge design or historical significance.

Fashion Consciousness

Psychographic segmentation considers consumer attitudes toward fashion and textiles. Some individuals are trend-conscious and seek the latest designs, while others may prioritize timeless, classic pieces. Understanding these preferences helps businesses align their product offerings with consumer expectations.

Innovation Adoption

Innovations in textile technology, such as smart textiles, may appeal to early adopters who embrace new trends and technologies. Psychographic segmentation identifies consumer segments more likely to adopt and appreciate cutting-edge textile innovations.

Fifthly, Distribution Channel

Online versus Vrick-And-Mortar

The choice of distribution channel is a critical factor in reaching the target audience. Online platforms cater to a global market and may attract a tech-savvy, digitally engaged audience. In contrast, brick-and-mortar stores may focus on local markets, offering a tangible and immersive shopping experience.

B2B Versus B2C

Businesses may cater primarily to consumers (B2C) or engage in business-to-business (B2B) transactions, supplying textiles for other industries. B2B-focused businesses may prioritize factors like bulk orders, customization for specific industries, and technical specifications.

Sixthly, Industry-Specific Segmentation

Fashion Industry

Within the fashion industry, segmentation may occur based on the type of clothing (e.g., casual wear, activewear, formalwear) or fashion subcultures (e.g., streetwear, bohemian fashion). Designers and brands tailor their textiles to meet the specific needs of these segments.

Home Textiles

Home textiles encompass a range of products, from bedding to curtains. Segmentation in this sector considers factors like design preferences, durability, and ease of maintenance, with different textiles catering to varied home decor styles.

Industrial Textiles

Industrial applications, such as automotive textiles or technical textiles for specialized industries, represent a distinct market segment. These textiles often prioritize functionality, durability, and specific technical requirements.

Art and Installations

Textiles used for artistic purposes, installations, and exhibitions form a unique segment. Artists and designers creating large-scale installations may have different requirements in terms of fabric characteristics and customization.

Market segmentation in the creative and value-added textile sector is a multifaceted strategy that considers a range of factors, from consumer demographics to product characteristics and industry-specific applications. Understanding these segments allows businesses to tailor their offerings, marketing efforts, and distribution strategies, ensuring they resonate with the diverse needs and preferences of their target audience. As the industry continues to evolve, effective segmentation will be essential for businesses seeking to navigate the complex and dynamic landscape of creative textiles.

C. **Supply Chain Analysis in the Creative and Value-Added Textile Sector**

The creative and value-added textile sector operates within a complex and multifaceted supply chain, encompassing various stages from raw material sourcing to the delivery of finished products to consumers. This supply chain involves a blend of traditional craftsmanship, global manufacturing, design innovation, and distribution channels. A comprehensive analysis of the supply chain in this sector is essential for businesses to optimize processes, ensure quality, and respond effectively to the dynamic demands of the market.

Raw Material Sourcing

The supply chain in the creative textiles sector typically begins with raw material sourcing, a critical stage that significantly influences the characteristics and sustainability of the final products. Different types of textiles require diverse raw materials, ranging from natural fibres like cotton and silk to synthetic fibres such as polyester. This stage involves a global network of suppliers and growers.

For example, the cultivation of organic cotton for sustainable textile production involves partnerships with certified organic farms. These farms adhere to strict standards, ensuring that the cotton is grown without synthetic pesticides or genetically modified organisms (GMOs) [3]. The sourcing of sustainable materials aligns with the growing consumer demand for eco-friendly products in the textiles sector.

Traditional Craftsmanship

Traditional craftsmanship remains a significant aspect of the creative textiles sector. Skilled artisans engage in manual processes such as weaving, dyeing, and embroidery, preserving cultural heritage and creating unique, handcrafted textiles. For instance, the handloom weavers in Varanasi, India, contribute to the production of Banarasi silk sarees, showcasing traditional techniques that have been passed down through generations [49].

Industrial Manufacturing

On the other hand, industrial manufacturing hubs, often situated in countries with lower labour costs, focus on mass production. Factories equipped with advanced machinery and technology engage in processes such as spinning, weaving, and dyeing on a large scale. China, with its vast textile industry, is a prominent example of a country contributing significantly to industrial manufacturing in the textile sector [34].

Design and Innovation

Once textiles are manufactured, they move to design studios where creative professionals conceptualize and design patterns, prints, and textures. Textile designers collaborate with fashion houses, brands, and manufacturers to create innovative designs that align with consumer trends. For example, the iconic Liberty London is renowned for its distinctive patterns and prints, showcasing the fusion of traditional craftsmanship with contemporary aesthetics [40].

Innovation Centres

Innovation centres play a crucial role in driving advancements in textile technology. These centres, often associated with research institutions or forward-thinking companies, focus on developing new materials and integrating technology into textiles. The

Fabricant, a digital fashion house, exemplifies the intersection of technology and textiles, creating digital-only garments through innovative design processes [66].

Quality Assurance in Manufacturing

Throughout the manufacturing process, quality control measures are implemented to ensure that textiles meet specified standards. Manufacturers conduct rigorous testing to identify defects, maintain colour consistency, and adhere to design specifications. Third-party testing laboratories, such as SGS and Bureau Veritas, are often engaged to ensure impartial evaluation and certification of textile quality [61].

Compliance and Regulations

Compliance with international and regional regulations is a critical aspect of quality control in the textile sector. Regulations cover product safety, environmental standards, and labour practices. For instance, the Global Organic Textile Standard (GOTS) sets stringent criteria for the environmental and social aspects of textile production, ensuring adherence to sustainable practices [29]. Global Organic Textile Standard (GOTS) is a private standard. It is managed by the International Working Group on Global Organic Textile Standard (IWG), which is a coalition of leading organizations in the organic textile industry. These organizations include the Organic Trade Association (OTA), the Japan Organic Cotton Association (JOCA), the International Association Natural Textile Industry (IVN), and the Soil Association.

GOTS was developed by these organizations to establish a unified global standard for organic textiles, covering both environmental and social criteria. While it is not a government-regulated standard, GOTS has gained widespread recognition and acceptance in the industry as a reliable certification for organic textiles. Manufacturers, brands, and retailers voluntarily seek GOTS certification to demonstrate their commitment to producing and selling organic, environmentally friendly, and socially responsible textiles.

The Global Organic Textile Standard (GOTS) is a leading certification for organic textiles worldwide. It sets stringent criteria for organic fibres' cultivation, environmentally and socially responsible manufacturing processes, and labelling requirements. GOTS ensures that textiles labelled as "organic" meet strict standards throughout the supply chain, from harvesting raw materials to labelling finished products.

This certification prohibits the use of toxic chemicals, genetically modified organisms (GMOs), and harmful dyes in textile production, promoting sustainability and environmental protection. Additionally, GOTS mandates fair labour practices, ensuring safe working conditions and fair wages for workers involved in the production of organic textiles. As consumer demand for sustainable and eco-friendly products grows, GOTS certification provides assurance of the authenticity and integrity of organic textiles in the global market.

Industry standards and government regulations play distinct but complementary roles in ensuring the quality, safety, and sustainability of textile products. Industry standards, such as the Global Organic Textile Standard (GOTS) and the American Association of Textile Chemists and Colourists (AATCC) standards, are developed by private organizations or industry associations to establish best practices, specifications, and guidelines for textile production and quality control. These standards are voluntary and are often adopted by manufacturers, brands, and retailers to demonstrate compliance with industry norms and consumer expectations.

On the other hand, government regulations are laws and mandates established by governmental bodies to enforce minimum requirements, safety standards, and labelling requirements for textile products. These regulations are legally binding and mandatory for manufacturers and importers, ensuring compliance with health, safety, and environmental standards set by the government. Examples include the Consumer Product Safety Improvement Act (CPSIA) in the United States and the EU Textile Regulation in the European Union.

One key difference between industry standards and government regulations lies in their scope and enforceability. Industry standards are typically more flexible and adaptable, allowing for innovation and continuous improvement in textile manufacturing practices. They provide guidance on best practices but do not carry legal consequences for non-compliance. In contrast, government regulations are rigid and enforceable by law, with penalties for non-compliance ranging from fines to product recalls or even legal action.

Another difference is in the development process and stakeholder involvement. Industry standards are often developed collaboratively by industry stakeholders, including manufacturers, trade associations, and consumer advocacy groups, ensuring that they reflect industry consensus and best practices. Government regulations, on the other hand, are developed and enforced by regulatory agencies and policymakers, often with input from industry stakeholders but ultimately driven by public health, safety, and consumer protection objectives.

Global Supply Chains

Distribution involves the transportation of textiles from manufacturing centres to global markets. The supply chain includes the movement of raw materials, semi-finished goods, and finished products. Global supply chains are intricate, involving coordination between manufacturers, logistics providers, and distributors to meet the demand from diverse geographical locations [24].

E-commerce Platforms

The rise of e-commerce has transformed the distribution landscape in the textiles sector. E-commerce platforms facilitate direct-to-consumer sales, allowing businesses to reach a global audience. Efficient logistics and supply chain management are essential for timely delivery, and partnerships with logistics companies, such as DHL or UPS, are common to ensure smooth operations [15].

Brick-And-Mortar Retail

Traditional retail channels remain relevant in the distribution of textiles. Textiles move through a network of distributors, wholesalers, and retailers before reaching the end consumer. Effective logistics play a crucial role in managing inventory and responding to seasonal demand fluctuations. Retailers like Zara have perfected a fast-fashion model, relying on a sophisticated logistics network to quickly deliver new designs to stores [70].

Fashion Retailers

Fashion retailers, both online and offline, are key players in the supply chain. They curate and display textiles for consumers, influencing trends and providing access to a wide range of products. Collaborating with retailers is essential for textile businesses seeking to reach a broader audience. High-end fashion retailers like Chanel or H&M, known for their distinct market positions, illustrate the role of retailers in shaping consumer choices [9].

Direct-To-Consumer Brands

Direct-to-consumer (DTC) brands have gained prominence in recent years, leveraging e-commerce to connect with consumers directly. Bypassing traditional retail channels allows DTC brands to maintain greater control over branding, pricing, and customer relationships. Companies like Ever Lane exemplify this model, emphasizing transparency in pricing and sourcing [62].

Repair and Upcycling

Aftermarket services contribute to sustainability efforts in the textiles sector. Some businesses offer repair and upcycling services, encouraging consumers to extend the lifespan of their textiles. Patagonia, a company known for its commitment to sustainability, promotes the repair of its products through the Worn Wear program, reducing the environmental impact of textile disposal [54].

Recycling Programs

Recycling programs address the challenge of textile waste. Companies, such as The North Face with its Clothes the Loop initiative, collect used textiles for recycling into new products. These programs contribute to a circular economy, minimizing the environmental footprint of textile production [67].

Data Analytics and Forecasting

Data analytics and forecasting play a crucial role in supply chain management. Businesses use data-driven insights to analyse consumer trends, optimize inventory levels, and respond to demand fluctuations. Predictive analytics tools help in anticipating market trends and adjusting production accordingly [17].

Blockchain Technology

Blockchain technology is emerging as a potential solution for enhancing transparency and traceability in the textile supply chain. By creating an immutable and secure ledger, blockchain can track the journey of textiles from raw material to finished product, ensuring authenticity and ethical sourcing. This technology addresses concerns related to counterfeiting and unethical practices [20].

Challenges and Opportunities

Challenges

The creative textile sector faces several challenges in its supply chain. Ethical concerns related to labour practices, environmental impact, and transparency remain critical issues. Rapid changes in consumer preferences and trends pose challenges in terms of forecasting and adapting to evolving market demands [18].

Opportunities

Amid challenges, opportunities arise for businesses in the creative textiles sector. Embracing sustainability, innovation, and technology presents avenues for differentiation. Sustainable practices, such as using eco-friendly materials and adopting circular economy principles, can appeal to environmentally conscious consumers. Moreover, leveraging technology for efficient supply chain management and adopting innovative business models contribute to long-term success [39].

The supply chain in the creative and value-added textile sector is a complex and interconnected network that spans various stages, from raw material sourcing to consumer delivery. Each stage involves diverse processes, stakeholders, and challenges. By understanding and optimizing each element of the supply chain, businesses in the creative textiles sector can enhance efficiency, ensure quality, and navigate the evolving landscape of the global textiles industry. As sustainability and innovation become central themes, the sector's supply chain will continue to evolve, presenting new opportunities and challenges for businesses to address in the pursuit of excellence.

In recent years, China has undergone a significant transition in its manufacturing sector, moving towards higher value-added production and increased automation. This shift has had a profound impact on the textile industry, which has long been a cornerstone of China's manufacturing prowess. Traditionally, China has been known as the world's leading exporter of textiles and garments, leveraging its abundant labour force and low production costs to dominate global markets. However, rising labour costs, changing consumer preferences, and technological advancements have spurred China to upgrade its manufacturing capabilities towards higher value-added products and automation.

One key driver of this transition is the Chinese government's initiatives to upgrade the manufacturing sector through policies such as "Made in China 2025." This strategic plan aims to transform China into a global leader in advanced manufacturing industries, including robotics, artificial intelligence, and high-tech textiles [51]. As a result, Chinese textile manufacturers are investing in advanced machinery and automation technologies to improve productivity, reduce labour costs, and enhance product quality.

Furthermore, China's textile industry is facing increasing competition from emerging manufacturing hubs in Southeast Asia, where labour costs remain lower. To remain competitive, Chinese textile manufacturers are adopting automation to streamline production processes and maintain cost competitiveness. Automation technologies such as robotic sewing machines, computerized knitting systems, and automated material handling systems are being deployed to increase efficiency and flexibility in textile production [69].

However, while automation offers opportunities for efficiency gains and cost savings, it also presents challenges for the labour force. The adoption of automation in the textile industry has led to concerns about job displacement and the future of traditional manufacturing jobs in China. To address these challenges, the Chinese government is implementing policies to support workforce training and skill development in advanced manufacturing technologies [75].

D. **Microeconomic Overview of the Creative and Value-Added Textile Sector**

The microeconomic landscape of the creative and value-added textile sector involves a closer examination of individual firms, market structures, pricing strategies, and factors influencing the supply and demand for textile products. This overview explores key microeconomic elements that shape the behaviour of businesses within the sector.

Market Structure

The creative and value-added textile sector exhibits characteristics of an oligopolistic market structure, where a small number of large firms dominate the industry. Major fashion houses, textile manufacturers, and retailers often wield significant market power. This concentration can lead to strategic interactions among key players, influencing pricing decisions and overall market dynamics [5].

Supply and Demand Dynamics

The microeconomic forces of supply and demand play a crucial role in determining the prices and quantities of textiles. Factors such as consumer preferences, fashion trends, and disposable income influence demand. On the supply side, the availability of raw materials, production costs, and technological advancements impact the quantity of textiles produced. The interplay between supply and demand influences pricing strategies adopted by businesses in the sector [42].

Elasticity of Demand

Consumer responsiveness to changes in price, known as elasticity of demand, is a significant factor in the creative and value-added textile sector. While certain basic textiles may exhibit inelastic demand (less responsive to price changes), high-end and fashion-forward products may have more elastic demand. The sector's firms must consider these elasticities when setting prices and adapting to shifts in consumer preferences [71].

Cost Structures

Understanding cost structures is essential for businesses in the creative and value-added textile sector. Production costs encompass raw materials, labour, technology, and marketing expenses. The sector often faces challenges related to labour-intensive processes, especially in traditional craftsmanship. Efficient cost management and strategic sourcing contribute to a firm's competitiveness in the market [7].

Pricing Strategies

Pricing strategies in the sector are diverse, ranging from premium pricing for high-end designer products to value-based pricing for more accessible items. Luxury brands often employ a skimming strategy, setting high initial prices to capture the market's top segment. Fast-fashion retailers, on the other hand, may adopt a penetration pricing strategy, offering lower prices to gain a larger market share. Dynamic pricing based on real-time demand and supply considerations is also becoming more prevalent, especially in online retail [50].

Innovation and Product Differentiation

Microeconomic decisions within firms often revolve around innovation and product differentiation. The creative and value-added textile sector thrives on introducing novel designs, materials, and manufacturing processes. Successful differentiation allows firms to command higher prices, enhance brand loyalty, and gain a competitive edge in the market. Firms invest in research and development to stay at the forefront of design and technology [57].

Market Entry Barriers

The sector presents both opportunities and challenges for new entrants. While creativity and innovation can be a source of competitive advantage, the presence of established brands and the need for significant initial investments create barriers to entry. Brand loyalty, distribution networks, and economies of scale enjoyed by large firms act as deterrents for smaller businesses entering the market [68].

Consumer Behaviour and Preferences

Microeconomic analysis in the creative and value-added textile sector delves into understanding consumer behaviour and preferences. Fashion cycles, trends, and the influence of social media play a pivotal role in shaping consumer choices. Firms conduct market research to anticipate shifts in preferences, allowing them to align their product offerings with current and emerging trends [63].

Profit Maximization Versus Corporate Social Responsibility (CSR)

Microeconomic decisions within firms often involve a trade-off between profit maximization and corporate social responsibility (CSR). As consumer awareness of ethical and sustainable practices grows, firms in the creative textiles sector must weigh the financial implications of adopting eco-friendly materials, fair labour practices, and sustainable production processes. Balancing profitability with CSR initiatives is a microeconomic challenge faced by many businesses in the sector [14].

Government Regulations and Policies

Microeconomic decision-making is influenced by government regulations and policies. In the textile sector, regulations cover aspects such as environmental standards, labour practices, and intellectual property rights. Compliance with these regulations is not only a legal requirement but also a strategic consideration for firms aiming to maintain a positive public image and meet evolving consumer expectations [65].

Competitive Rivalry

Microeconomic analysis considers the level of competitive rivalry within the sector. Firms compete for market share through aggressive marketing, product innovation, and efficient supply chain management. Understanding the competitive landscape helps businesses make informed decisions about pricing, promotion, and product development [56].

The microeconomic overview of the creative and value-added textile sector provides insights into the intricacies of individual firms' decision-making processes within the broader market structure. The interplay of supply and demand, pricing strategies, innovation, and considerations of cost and profit contribute to the dynamic nature of the sector. Businesses navigating this landscape must continually adapt to changing consumer preferences, technological advancements, and global market dynamics to remain competitive and sustainable.

E. **Impact of Creative and Value-Added Textiles in Global Economy**

The creative and value-added textile sector has a profound influence on the global economy, contributing significantly to various aspects such as trade, employment, innovation, and cultural expression. This industry, which encompasses the production of high-quality textiles, innovative designs, and artisanal craftsmanship, plays a crucial role in shaping economic dynamics on a global scale.

Contribution to Global Trade

The creative and value-added textile sector is a major player in international trade. Textile products, ranging from high-end fashion items to unique artisanal crEAtions, are traded across borders, contributing to the global flow of goods. According to the World Trade Organization (WTO), textiles and apparel represent a substantial share of world merchandise trade, with exports reaching hundreds of billions of dollars annually [72]. The sector's contribution to trade enhances economic interconnectedness and provides opportunities for countries to specialize in different segments of the value chain.

Employment Generation

One of the sector's significant impacts is its role as a generator of employment worldwide. From skilled artisans practising traditional craftsmanship to workers in modern manufacturing facilities, the creative and value-added textile sector employs millions of people globally. Particularly in developing countries, the industry provides valuable job opportunities, contributing to poverty alleviation and socio-economic development. The International Labour Organization (ILO) highlights the sector's role in supporting livelihoods and fostering economic growth through employment (ILO, 2021).

Cultural Expression and Heritage

The creative and value-added textile sector often draws on cultural heritage and traditional craftsmanship, contributing to the preservation and promotion of cultural diversity. Artisanal practices, unique designs, and traditional techniques are embedded in the textiles produced by this sector. For example, regions known for their distinctive textile traditions, such as Bangladesh's handwoven silk sarees or Scotland's tartan patterns, showcase the intersection of creativity, culture, and economic activity. The preservation of cultural expression through textiles adds a qualitative dimension to the sector's global impact.

Innovation and Design Excellence

Innovation is a hallmark of the creative and value-added textile sector. The industry continually pushes boundaries in design, materials, and manufacturing processes. High-end fashion houses invest heavily in research and development to introduce cutting-edge designs and incorporate new technologies. The intersection of art and technology has given rise to smart textiles, 3D-printed fabrics, and sustainable innovations. This commitment to innovation not only propels the sector forward but also fosters a culture of continuous improvement and creativity [23].

Economic Value Chain

The sector operates within a complex economic value chain that spans multiple stages, from raw material sourcing to retail. The value chain involves diverse stakeholders, including cotton farmers, textile manufacturers, designers, retailers, and consumers. Each stage contributes to the overall economic impact, with value addition occurring at various points. For example, the incorporation of unique designs, high-quality materials, and sustainable practices enhances the value of the final product, creating economic value along the entire chain [31].

Sustainable Practices and Corporate Social Responsibility (CSR)

In recent years, there has been a growing emphasis on sustainability and corporate social responsibility (CSR) within the creative and value-added textile sector. Consumers, investors, and regulatory bodies increasingly demand environmentally responsible practices. Many companies in the sector are adopting sustainable sourcing of materials, eco-friendly production processes, and ethical labour practices. Initiatives such as fair-trade certifications and transparency in the supply chain contribute to the sector's positive global economic impact [52].

Economic Diplomacy and Soft Power

The creative and value-added textile sector serves as a tool for economic diplomacy and the projection of soft power. Countries with strong textile traditions leverage their cultural and artistic capital to enhance their global image. Luxury brands associated with certain regions contribute not only to the economic prosperity of those regions but also to the perception of the country's craftsmanship and style. For example,

Italian fashion houses are renowned for their craftsmanship, contributing to Italy's image as a hub of luxury and creativity [35].

Consumer Spending and Lifestyle

Consumer spending on textiles and apparel is a key driver of economic activity globally. The creative and value-added textile sector contributes to the lifestyle choices of consumers, influencing trends and preferences. High-quality, unique, and aesthetically pleasing textiles often command premium prices, leading to increased consumer spending. The sector's ability to tap into consumer desires for individuality and expression contributes to its economic significance [8].

Global Supply Chains and Resilience

The sector's global impact is also reflected in its participation in complex and interconnected supply chains. Raw materials sourced from one region may undergo manufacturing in another, and the final product may be sold in diverse markets. The COVID-19 pandemic highlighted the vulnerabilities in global supply chains, leading to disruptions. However, the sector's adaptability and resilience were evident as businesses adjusted production, diversified suppliers, and explored new ways of reaching consumers [47].

The global economic impact of the creative and value-added textile sector is multifaceted, encompassing trade, employment, cultural expression, innovation, and sustainability. As a dynamic and interconnected industry, it not only contributes to economic growth but also shapes consumer lifestyles and perceptions of cultural identity. With a focus on sustainability and ethical practices, the sector has the potential to lead the way in responsible and impactful economic contributions on a global scale.

F. The Financial Impact of Creative and Value-Added Textiles Globally

The creative and value-added textile sector exerts a substantial financial impact on the global economy, encompassing a diverse range of activities from traditional craftsmanship to cutting-edge design. This sector's economic significance is evident in its contribution to various financial aspects, including trade, revenue generation, employment, and investment. This discussion explores the financial dynamics of the creative and value-added textile sector on a global scale.

Economic Contribution and Revenue Generation

The creative and value-added textile sector is a significant contributor to global economic output. Revenue generation in this sector involves a spectrum of activities, including the production of high-quality fabrics, innovative designs, and artisanal creations. According to market reports, the global textile and apparel market size is in the trillions of dollars, reflecting the sector's economic magnitude [64]. Revenue generation occurs not only through product sales but also through licensing, collaborations, and brand partnerships, further amplifying the financial impact.

Trade and Export Earnings

The sector plays a crucial role in global trade, with countries engaging in the export and import of textiles and apparel products. Many nations have a comparative advantage in specific segments of the value chain, leading to a complex web of international trade relationships. The financial impact is evident in export earnings, trade balances, and the creation of trade surpluses for textile-exporting countries. For instance, countries like China and Bangladesh are among the top textile exporters globally, contributing significantly to their foreign exchange reserves [72] (Table 1).

Employment and Wage Generation

The financial impact of the creative and value-added textile sector extends to employment and wage generation. The labour-intensive nature of textile manufacturing and craftsmanship provides jobs for millions of people globally. Employment, in turn, leads to wage generation, contributing to individual incomes and household livelihoods. This financial impact is particularly pronounced in countries where the textile industry serves as a major source of employment and economic stability. It aligns with the sector's role in poverty reduction and socio-economic development [36].

Investment and Innovation

Investment in the creative and value-added textile sector fuels innovation, technology adoption, and infrastructure development. Companies invest in research and development to stay at the forefront of design and sustainability. The financial impact of such investments is twofold: it enhances the competitiveness of individual firms and stimulates economic growth through technological advancements. Governments and private entities invest in initiatives that support the sector's evolution, contributing to the overall financial vibrancy of the industry [23].

Market Capitalization of Fashion Brands

The financial impact of the sector is evident in the market capitalization of leading fashion brands. Luxury fashion houses, known for their creative designs and value-added products, command substantial market value. For example, reports on the market capitalization of major luxury fashion brands highlight the economic significance of these companies within the broader fashion and textile sector. The financial

Table 1 Growth rate of world GDP and export

	World GDP	Total merchandise trade	Textiles export	Clothing export
2020	-3.3%	-7.2%	7.5%	-9.1%
2021	5.8%	26.5%	7.8%	21.9%
2022	3.3%	3.5%	3.1%	3.4%

Textiles and clothing exports in 2022 were estimated based on data from WTO (2022) and World Bank (2022)

Chart: by Dr. Sheng Lu • Created with Datawrapper

success of these brands is often linked to their ability to create and market unique, high-value products [13] (Fig. 1).

Consumer Spending and Retail Revenue

Consumer spending on textiles and apparel is a key driver of the financial impact of the sector. The demand for creative and value-added textiles influences consumer purchasing behaviour. High-end and designer products often command premium prices, contributing to retail revenue. The financial success of retail businesses within the sector is closely tied to their ability to meet consumer expectations, adapt to trends, and create a desirable brand image. Retail revenue, in turn, supports the entire value chain, from manufacturers to designers [46] (Table 2).

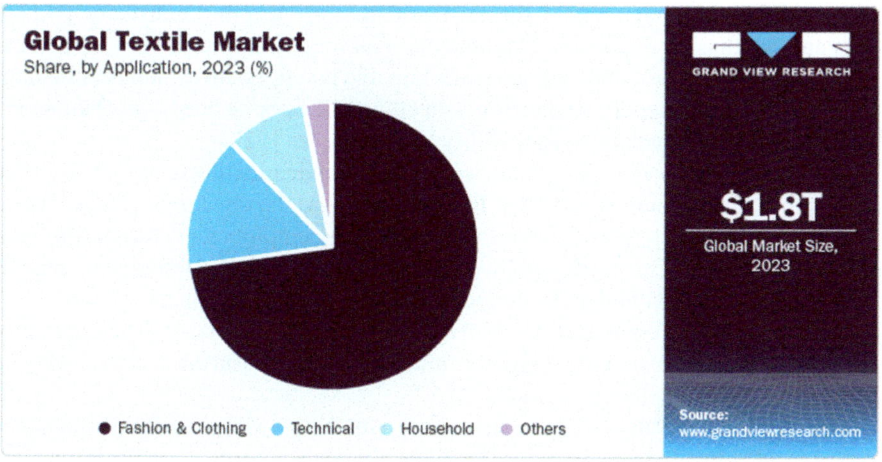

Fig. 1 Global textile market

Table 2 Regional textile trade patterns

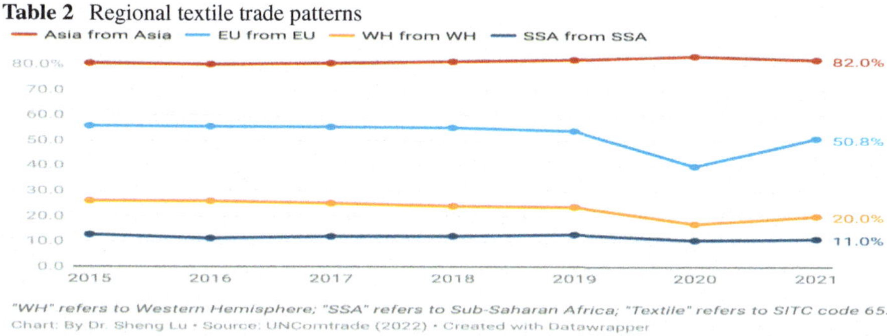

Financial Resilience and Adaptability

The financial resilience and adaptability of the creative and value-added textile sector were tested during global crises, such as the COVID-19 pandemic. The ability of businesses to navigate disruptions, reassess supply chains, and adapt to changing consumer behaviours demonstrates the financial agility of the sector. The adoption of e-commerce, direct-to-consumer models, and digital marketing strategies reflects the industry's commitment to financial resilience and innovation in response to external challenges [47].

Financial Challenges and Sustainable Practices

While the financial impact of the sector is significant, it also faces challenges that have financial implications. Fast fashion, overconsumption, and environmental concerns have led to increased scrutiny, potentially impacting consumer perceptions and purchasing decisions. The financial viability of businesses in the sector is increasingly tied to their ability to adopt sustainable practices. Investments in sustainable sourcing, eco-friendly production, and ethical labour practices contribute to the sector's long-term financial sustainability and resilience [52].

The financial impact of the creative and value-added textile sector on a global scale is multifaceted and substantial. It extends beyond the direct revenue generated through product sales to encompass trade balances, employment, innovation, and investments. The sector's adaptability to changing consumer preferences, market dynamics, and global challenges underscores its financial resilience.

As the industry continues to evolve, the financial decisions and strategies employed by businesses will shape its future trajectory, ensuring a continued and impactful presence in the global economy.

The creative and value-added textile sector stands as a vibrant and indispensable force on the global stage, weaving together elements of culture, innovation, and economic dynamism. From the intricacies of traditional craftsmanship to the forefront of technological advancements, this sector resonates far beyond the production of fabrics—it shapes the very fabric of our interconnected world. The significance of this industry manifests across several dimensions, affirming its role as a driver of economic growth, a cultural ambassador, and a trendsetter in sustainable practices.

In conclusion, the global importance of the creative and value-added textile sector is multi-faceted and dynamic. Its economic impact reverberates through trade balances, consumer spending, and market capitalization. As a cultural custodian, it preserves and disseminates diverse traditions. As an innovation hub, it continually shapes the future of design and materials. As a sustainable torchbearer, it leads the charge towards responsible business practices. And as an employment generator, it transforms lives and communities.

As we look forward, the sector's importance is poised to grow further. Navigating the challenges of fast fashion, technological disruptions, and global uncertainties will require resilience and innovation. Sustainable practices will become not just a choice but a necessity. The sector's ability to balance economic imperatives with cultural preservation, innovation with tradition, and global ambitions with local sensitivities will determine its enduring impact on the world stage. The creative and value-added textile sector, with its rich tapestry of contributions, remains a testament to the confluence of art, commerce, and culture, weaving a legacy that extends far beyond the threads of fabric.

The Global Textile Market is anticipated to rise at a considerable rate during the forecast period, between 2024 and 2031. In 2022, the market is growing at a steady rate and with the rising adoption of strategies by key players, the market is expected to rise over the projected horizon.

The global textile market is divided based on application, end user, and region, with a specific focus on manufacturers situated in various geographic areas. The study offers a comprehensive analysis of diverse factors that contribute to the industry's growth. It also outlines potential future impacts on the industry through various segments and applications. The report includes a detailed pricing analysis for different types, manufacturers, regional considerations, and pricing trends.

The Textile Share report delivers an overview of the market's value structure, cost determinants, and key driving factors. It assesses the industry landscape and subsequently examines the global landscape encompassing industry size, demand, applications, revenue, products, regions, and segments. Moreover, the Textile Market report presents the competitive scenario in the market among distributors and manufacturers, encompassing market value assessment and a breakdown of the cost chain structure.

Global Textile Market Report: Key Insights (Industry Oriented Tactics 2023)

Regional Share

The Textile Market report provides market size data for various regions, including North America, Europe, Asia Pacific, Latin America, the Middle East, and Africa. In 2022, North America dominated the Textile market, followed by Europe, while Asia Pacific held a significant share.

Segment Overview

The market is segmented based on Type and Application. In 2022, Down accounted for a notable percentage of the textile market, with Feather also contributing significantly. Comforters held a significant share as well, and apparel played a role in the market.

Competitive Landscape

The report offers a comprehensive analysis of prominent players who have substantial market shares. It includes information on the concentration ratio and provides detailed insights into the market performance of each player. This allows readers to gain a holistic understanding of the competitive landscape and better knowledge of their competitors.

Textile Market Report Detailed Overview

The global textile market size was valued at USD 1,206,527.99 million in 2021 and is expected to expand at a CAGR of 3.93% of 3.93% during the forecast period, reaching USD 1,520,579.14 million by 2027.

Textiles are flexible materials composed of natural or man-made fibre networks.

The report combines extensive quantitative analysis and exhaustive qualitative analysis, ranges from a macro-overview of the total market size, industry chain, and market dynamics to micro details of segment markets by type, application and region, and, as a result, provides a holistic view of, as well as a deep insight into the Textile market covering all its essential aspects.

For the competitive landscape, the report also introduces players in the industry from the perspective of the market share, concentration ratio, etc. and describes the leading companies in detail, with which the readers can get a better idea of their competitors and acquire an in-depth understanding of the competitive situation. Further, mergers and acquisitions, emerging market trends, the impact of COVID-19, and regional conflicts will all be considered.

References

1. Ashton TS (2007) The industrial revolution, 1760–1830. Oxford University Press
2. Austin G (2008) The 'reversal of fortune' thesis and the compression of history: perspectives from African and comparative economic history. J Int Dev 20(8):996–1027
3. Balfour K (2009) Organic cotton: from field to final product. Text Exch. Accessed https://textileexchange.org/
4. Baumeister RF, Vohs KD, Funder DC (2007) Psychology as the science of self-reports and finger movements: whatever happened to actual behavior? Perspect Psychol Sci 2(4):396–403
5. Baumol WJ, Blinder AS (2015) Microeconomics: principles and policy. Cengage Learning
6. Bender B (1990) Gatherer-hunter to farmer: a social perspective. World Archaeol 21(2):124–136
7. Besanko D, Braeutigam R, Gibbs M (2013) Microeconomics. Wiley
8. Bhardwaj V, Fairhurst A (2010) Fast fashion: response to changes in the fashion industry. Int Rev Retail Distrib Consum Res 20(1):165–173
9. Bruce M, Daly L (2006) Retailing in the 21st century: current and future trends. Springer
10. Brynjolfsson E, McAfee A (2014) The second machine age: work, progress, and prosperity in a time of brilliant technologies. W. W. Norton & Company
11. Bulmer-Thomas V (2003) The economic history of Latin America since independence. Cambridge University Press
12. Burnham DK (1980) Warp and weft: a textile terminology. Royal Ontario Museum
13. Business of Fashion (2021) Global fashion index. Accessed https://www.businessoffashion.com/
14. Carroll AB (1999) Corporate social responsibility: evolution of a definitional construct. Bus Soc 38(3):268–295
15. Carter CR, Rogers DS (2008) A framework of sustainable supply chain management: moving toward new theory. Int J Phys Distrib Logist Manag 38(5):360–387
16. Chakrabarty D (1989) Rethinking working-class history: Bengal 1890–1940. Princeton University Press
17. Chopra S, Meindl P (2016) Supply chain management: strategy, planning, and operation. Pearson

18. Cohen MA, Roussel J (2005). Strategic supply chain management: the five core disciplines for top performance. McGraw-Hill
19. Cumming E, Kaplan W, Pritchard A (1993) The arts and crafts movement. Thames & Hudson
20. Dias T, Lima P, Oliveira T (2019) Blockchain in the supply chain: a systematic literature review. Int J Inf Manag 46:70–82
21. Diderot D, d'Alembert J (1751) Encyclopédie, ou Dictionnaire raisonné des sciences, des arts et des métiers
22. Falcon JB (1728) Description de la nouvelle machine pour broder les rubans, etc.
23. Fang H, Nian H (2019) Exploring the driving forces of innovation in textile and apparel industries: a dynamic capability view. Sustainability 11(20):5677
24. Fernie J, Sparks L (2014) Logistics and retail management: emerging issues and new challenges in the retail supply chain. Kogan Page Publishers
25. Fields GS (2004) Dualism in the labor market: a perspective on the lewis model after half a century. Manch Sch 72(6):724–735
26. Fletcher K, Grose L (2012) Fashion & sustainability: design for change. Laurence King Publishing.
27. Fletcher K (2017) Sustainable fashion and textiles: design journeys. Routledge
28. Fletcher K (2018) Craft of use: post-growth fashion. Routledge
29. GOTS (Global Organic Textile Standard) (n.d.) About us. Accessed https://www.global-standard.org/
30. Geng R, Qian W (2017) Sustainable fashion supply chain: lessons from H&M. Sustainability 9(12):2333
31. Gereffi G, Frederick S (2010) The global apparel value chain, trade and the crisis: challenges and opportunities for developing countries. United Nations Industrial Development Organization (UNIDO)
32. Gillow J, Barnard N (1991) Traditional Indian textiles. Thames & Hudson
33. Hane M (1992) Modern Japan: a historical survey, 3rd edn. Westview Press
34. Hill CWL (2009) International business: competing in the global marketplace. McGraw-Hill Education
35. Holt DB (2016) Brands and branding: research findings and future priorities. Marketing Science Institute
36. ILO (International Labour Organization) (2021) World employment and social outlook: trends 2021. Accessed https://www.ilo.org/
37. Kaplinsky R, Köhler G (2004) One thing leads to another: promoting industrialization by making linkages work. Inst Dev Stud Bull 35(3):40–49
38. Klein N (2002) No logo: taking aim at the brand bullies. Picador
39. Kotler P, Kartajaya H, Setiawan I (2019) Marketing 4.0: moving from traditional to digital. Wiley
40. Liberty (nd) Heritage. Accessed https://www.libertylondon.com/
41. Mackie L (1987) Ethnographic evidence for weaving as an ancient technology. Sci Am 256(4):112–120
42. Mankiw NG (2014) Principles of microeconomics. Cengage Learning
43. Marks R (1991) The origins of the modern world: a global and ecological narrative from the fifteenth to the twenty-first century. Rowman & Littlefield
44. McCloskey DN (2016) Bourgeois equality: how ideas, not capital or institutions. University of Chicago Press, Enriched the World
45. McCusker JJ, Menard RR (1985) The economy of British America, 1607–1789. UNC Press Books
46. McKinsey & Company (2020) COVID-19 and fashion: how the industry is responding. Accessed https://www.mckinsey.com/
47. McKinsey & Company (2021) The state of fashion 2021: in search of promise in perilous times. Accessed https://www.mckinsey.com/
48. Mokyr J (1999) The industrial revolution and the Netherlands: why did it not happen? J Econ Hist 59(1):3–20

49. Mukherjee S (2015) Artisanal silk weaving in Varanasi: challenges and solutions. Int J Sci Res (IJSR) 4(7):1972–1975
50. Nagle TT, Müller G (2017) The strategy and tactics of pricing: a guide to profitable decision making. Routledge
51. National Development and Reform Commission (2015) Made in China 2025. Accessed http://en.ndrc.gov.cn/newsrelease/201503/t20150330_669367.html
52. Niinimäki K, Peters G, Dahlbo H, Perry P, Rissanen T, Gwilt A (2020) The environmental price of fast fashion. Nat Rev Earth Environ 1(4):189–200
53. Parker Pearson M (1999) The archaeology of death and burial. Texas A&M University Press
54. Patagonia (n.d.) Worn wear. Accessed https://wornwear.patagonia.com/
55. Pevsner N (1936) An enquiry into industrial art in England. Cambridge University Press
56. Porter ME (1979) How competitive forces shape strategy. Harv Bus Rev 57(2):137–145
57. Porter ME (1985) Competitive advantage: creating and sustaining superior performance. Free Press
58. Quinn B (2014) Textile futures: fashion, design, and technology. Berg
59. Rodrik D (2017) Premature deindustrialization. J Econ Growth 22(1):1–33
60. Rutherford TF (2018) Sustainability: a cultural history. Oxford University Press
61. SGS (n.d.) Textile & apparel testing services. Accessed https://www.sgs.com/
62. Shah S (2019) Transparency in the fashion industry: a case study of Everlane. Accessed https://www.researchgate.net/
63. Solomon MR (2019) Consumer behavior: buying, having, and being. Pearson.
64. Statista (2021) Global apparel market size. Accessed https://www.statista.com/
65. Stigler GJ (1971) The theory of economic regulation. The Bell J Econ Manag Sci 2(1):3–21
66. The Fabricant (n.d.) Digital fashion house. Accessed https://www.thefabricant.com/
67. The North Face (n.d.) Clothes the loop. Accessed https://www.thenorthface.com/
68. Tirole J (1988) The theory of industrial organization. MIT Press
69. Tong X (2018) The impact of automation on employment: the case of China's textile and apparel industry. Asian Econ Policy Rev 13(2):292–310
70. Tukker A, Emmert S, Charter M, Vezzoli C, Sto E, Andersen MM (2016) Fostering change to sustainable consumption and production: an evidence-based view. J Clean Prod 132:88–103
71. Varian HR (2014) Intermediate microeconomics: a modern approach. W. W. Norton & Company.
72. WTO (2021) International trade statistics. World Trade Organization
73. Wild JP (2015) Textiles in archaeology. Shire Publications
74. Wilkins C (1990) The Jacquard loom and the invention of the computer. In: Jacquard's web. Oxford University Press, pp 1–19
75. Yuan S (2019) Will automation lead to a jobless future in China? A double perspective based on labor economics and employment psychology. Sustainability 11(14):3961

Historical and Contemporary Development of Textile Design

Fatima Iftikhar and Suniya Tariq

Abstract This chapter provides an in-depth exploration of design and techniques in textile design within the context of India, Pakistan, and Bangladesh—countries previously known as the Subcontinent. Drawing from existing literature, the chapter sheds light on the diverse techniques employed by artisans, designers, and developers, serving various purposes such as cultural activism, survival, and fostering a sense of social belonging. The craftsmanship in these nations is a remarkable testament to their rich heritage, with each country offering a plethora of distinct crafts and techniques unique to their respective regions. The chapter thoroughly explores the expansive world of textile craft and techniques, spanning the period from the 1900s to the year 2023. It culminates with concise predictions regarding the future of textiles, encompassing the design, methods, and processes that these countries are likely to contribute. This showcases the dynamic and ever-evolving nature of the textile industry, reflecting a blend of historical traditions and promising innovations in this captivating field.

1 Background

Textiles hold a profound place in the annals of human history, tracing their origins back to ancient times, with tangible evidence dating as far as 3400 BC and 5000 BCE. Woven fabrics have been discovered in the archives of civilizations like Egypt, Mesopotamia, and the Indus Valley. However, the precise moment when humanity first adorned clothing remains a subject of ongoing archaeological debate. Throughout history, diverse and advanced societies autonomously developed their unique textile patterns, employing specialized techniques and a diverse array of raw

F. Iftikhar (✉) · S. Tariq
Pakistan Institute of Fashion and Design, Lahore, Pakistan
e-mail: f.iftikhar@deakin.edu.au

Deakin University, Geelong, Australia

S. Tariq
e-mail: s.tariq@deakin.edu.au

H. Jamshaid and A. Dad (eds.), *Creative Textile Industry*, SDGs and Textiles,
https://doi.org/10.1007/978-981-97-3802-1_4

materials like hemp, cotton, wool, and silk. These remarkable civilizations achieved milestones across various domains, leaving a lasting legacy through the production of exquisitely designed textiles, each intricately woven to reflect their distinct cultural and geographical tapestries.

In the centuries preceding the nineteenth century, the Subcontinent had thrived as a prominent producer and supplier of textiles in various forms. Textile production was widespread throughout the subcontinent, catering to domestic consumption, the Indian Ocean, and the global market. While the majority of local demand was met through village-level production with minimal trade, the portion of output entering regional or sub-continental trade was far from negligible, particularly in luxury textiles marked by regional specialization. Notably, long-distance trade in textile raw materials played a vital role, exemplified by the Bengal cotton textile sector's reliance on cotton imports from Gujarat, which, in turn, depended on raw silk imports from Bengal for its silk textile sector [1].

In recent decades, textiles have emerged as a linchpin of global trade. The textile industry has undergone multiple phases of development, evolving from the cultivation of vital raw materials to the creation of distinctive textures and intricate designs, often involving labor-intensive crafting techniques such as batik, lehria, block printing, and ajrak.

Once recognized as the Subcontinent, this South Asian region has evolved into three prominent textile-producing nations: Pakistan, India, and Bangladesh. These nations boast abundant raw materials, a wealth of crafting techniques, and a deep reservoir of design inspiration, making significant contributions to the global textile trade, with cotton and silk standing as their top exports. Over the course of a millennium, textiles have served as conduits, opening up Southeast Asia to long-distance trade and nurturing economic exchanges on a global scale.

The allure of this region's textile industry captured the attention of European companies in the early sixteenth century. The Portuguese, trailblazers in discovering a sea route from Europe to the East, bypassing onerous taxes on Middle Eastern goods, embarked on this commercial venture. Subsequently, the British East India Company and the Dutch East India Company came into existence in 1600 and 1602, respectively. These entities procured textiles from India, bartering them for the spices cultivated in the Malay Islands, which were then traded across Europe and Asia. Indian textiles swiftly permeated European markets, earning widespread popularity, as evidenced by the numerous words of Indian origin that integrated into the English language, including calico, pajama, gingham, dungaree, chintz, and khaki. During the nineteenth century, India occupied a pivotal place as the 'jewel in the crown' of the British Empire, with Indian textiles cherished for their beauty and fashionable appeal. In 1851, textiles from India occupied a significant space at The Great Exhibition in the Crystal Palace in Hyde Park, London, highlighting their allure. Today, these luxurious textiles, revered for centuries, reside in museums, are meticulously studied, and are categorized based on their distinctive production patterns [2].

This exploration unearths the rich tapestry of textiles, unveiling their historical significance and tracing their journey from ancient epochs to the modern era. It provides insight into the captivating interplay of culture, trade, and design that has woven the enduring legacy of the textile industry.

2 Design

Design is an intricately woven creative process, encompassing the conception, planning, and execution of elemental arrangements to fulfill specific purposes or goals. Its reach extends across diverse domains, including graphic design, textiles, fashion, and more. As a discipline, design is rooted in the dynamic interplay between individuals, often referred to as 'users,' and the products they interact with. This process considers a multifaceted blend of aesthetic, functional, contextual, cultural, and societal factors. While design as a formalized discipline represents a relatively modern concept, its influence is far-reaching.

Central to the art of design are several foundational elements. These include creative problem-solving, which seeks innovative solutions; aesthetics, which shapes visual appeal; functionality, ensuring effective use; and user requirements, which prioritize user-friendly experiences. Additionally, design thrives on innovation, continually pushing boundaries and thinking outside the box. Importantly, design serves as a channel for conveying information, ideas, and emotions, and it dynamically adapts to evolving needs and contexts.

It's essential to recognize that the contemporary discipline of design has evolved over time. In its early years, not everything produced adhered to the structured design principles of today, with not everything being fundamentally developed for a specific end use, as is often the case in the twenty-first century. This evolution has underscored the dynamic nature of design as it continues to shape and be shaped by our ever-changing world [3].

3 Textile Design

Textile design, a multifaceted and dynamic field, encompasses process consists of lengths more than just visual implementation of design. It initiates at the stage of yarn to fabric construction, to dyeing to print to surface decoration and embellishment textile design process takes up each milestone with equal emphasis towards product ideation and development. These and various other techniques are implemented to produce fabrics that are not only functionally mature but also aesthetically pleasing. This intricate and versatile discipline revolves around the intricate interplay of design management and technology, which drives its development and evolution.

The roots of textile design stretch back to the dawn of human civilization, intertwined with ancient history. This craft is etched into the cultural tapestry of societies, with mentions found in poetry, myths, and cultural narratives from time immemorial. Textiles, in their various forms, have been instrumental in shaping human history.

They served as more than just functional materials; they held symbolic and cultural significance, bestowing stature upon their owners while seamlessly integrating into daily life [4].

The essence of textile design lies not only in its technical proficiency but also in its ability to merge the past with the present, threading together ancient traditions and contemporary innovations. It is a testament to human creativity, adaptability, and the enduring legacy of a craft that has been intertwined with our existence since time immemorial [5].

4 Evolution of Textile Design; India, Pakistan, and Bangladesh

The South Asian region has undergone significant transformation throughout the past century, encompassing not only textile construction but also design and decoration. Each country in this region has experienced substantial growth, contributing to the development of their distinctive craft identities as they stand today. Pakistan, India, and Bangladesh were once part of the same territory; they have evolved with shared similarities and notable differences, which together contribute to the unique and diverse character of their creative work. This evolution and divergence have enriched the cultural and artistic landscape of the South Asian region, shaping its rich tapestry of craftsmanship (Fig. 1).

4.1 Weave

Textile construction is an exquisite art, a process of creating textile surfaces using yarn, thread, and other raw materials. The production processes comprise weaving techniques such as weaving, knitting, crocheting, felting, braiding, etc. Up until the end of the nineteenth century, craftsmen were known to employ a variety of these techniques, mostly by hand, for the development of beautified surfaces.

Weaving has long been the fundamental textile production technique of the subcontinent region, currently comprising the countries of India, Pakistan, and Bangladesh. Weaving is a simple process of interlacing two distinct sets of yarns or threads known as tana (warf) and bana (weft) at a right angle, resulting in the formation of textile surfaces (fabric or cloth). Weaving, which has seeped into the roots of the South Asian region, has a vast history in woven textiles, having originated as far back as 2000 years ago [6]. Some of the textile's qualities have originated and been practiced in specified areas; quite a few have had a wide spread of practice over the Indian, Pakistan, and Bangladesh belts, being found in more than one city. Each city and region is known for distinct techniques and practices, such as Kanjeevaram silk from Kanchipuram, Tamil Nadu, Thari weave from Tharparker Sindh,

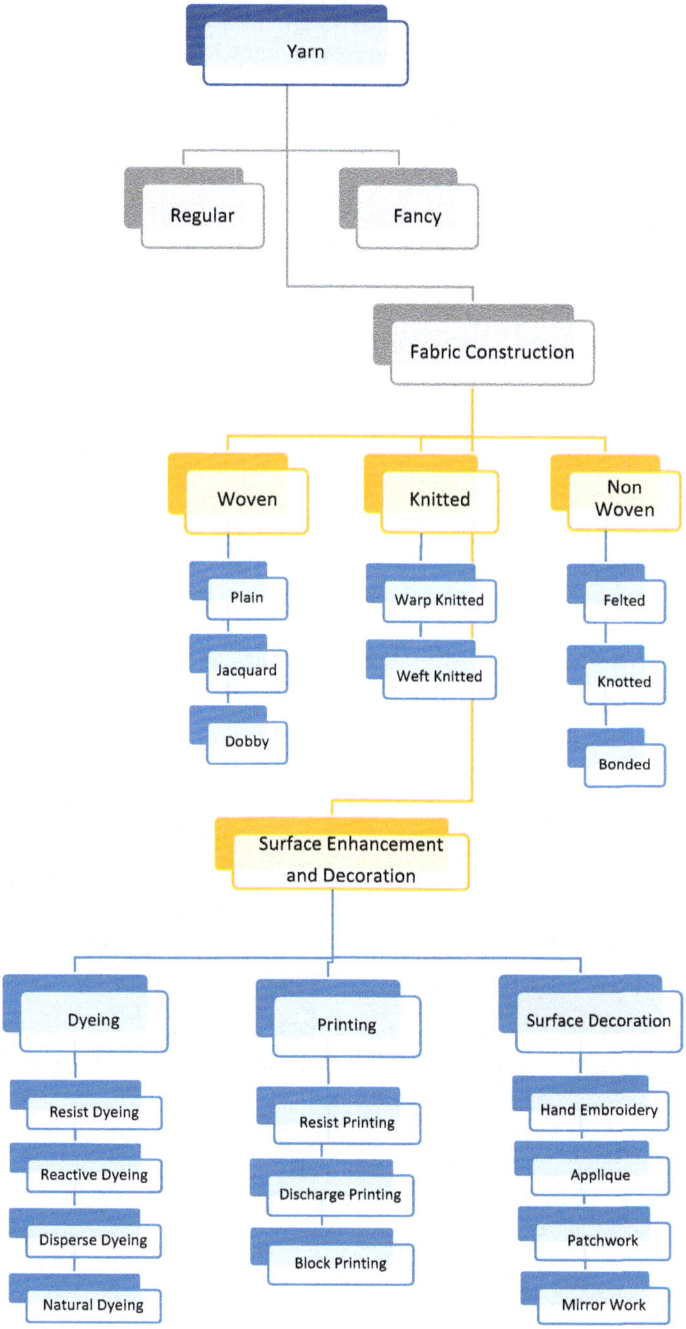

Fig. 1 Textile design, material, and processes (by Authors)

and Muslin from Dhaka. Having grasped design inspiration from varying areas, each craft has long been known to produce pieces unique in nature [7].

4.1.1 Banarasi Silk Sarees

The Banarasi saree, which has its origins in Varanasi, India, possesses a significant historical background that can be traced back to the Mughal era during the fourteenth century. The weavers hailing from Banaras were renowned for their skill in crafting elaborate silk brocades, embellished with strands of silver and gold zari. The art form originated in Vanaras and subsequently spread too many regions of India. The sarees in question are held in high esteem due to their elaborate weaving techniques and rich motifs. These motifs showcase themes inspired by the Mughal era, such as interwoven floral and foliate patterns, as well as decorative elements like kalga and bel. The technique encompasses intricate gold craftsmanship, tightly woven fabric, elaborate figurative representations, metallic visual enhancements, decorative borders, intricate lattice patterns, and enamel embellishments. Following the partition of British India in 1947, a number of proficient weavers hailing from Banaras and Delhi relocated to Pakistan and East Pakistan, presently known as Bangladesh. Workshops were created in large centres such as Karachi, Lahore, Khairpur, and Mirpur [8]. Through the transmission of their expertise across successive generations, individuals persist in safeguarding the aesthetic allure and cultural importance of their profession [9] (Fig. 2).

Fig. 2 Sample images of Banarasi silk sarees (by Authors)

Fig. 3 A Jamdani shawl in black (by Authors)

4.1.2 Jamdani

Jamdani cloth, a renowned textile from Bangladesh, has a rich history dating back over two thousand years. The name Jamdani comes from the Persian words "jam" (meaning floral) and "Dani," which refers to a vase or container named after decorative floral patterns in Dhakai Muslin. The name alone serves as a testament to the fabric's historical importance as a highly esteemed textile throughout the Mughal period [10]. The manufacturing process involves selecting premium-grade cotton yarn and a laborious dyeing process using natural or chemical dyes. Traditional handlooms are used to weave the cloth, allowing for intricate patterns. The design language of Jamdani is characterized by motifs like "panna hazaar," "kalka," "toradar," and "phulwar."

Today, production continues in over 200 villages in Dhaka. Jamdani fabric is widely acclaimed for its exquisite texture and remarkable quality, therefore perpetuating its esteemed position as a revered textile within the nation of Bangladesh (Fig. 3).

4.1.3 Khadi

The fabric known as Khadi, which is produced through the labor-intensive processes of hand-spinning and hand-weaving, possesses a rich historical heritage that can be traced back to the twelfth century A.D. Its importance was further amplified throughout India's struggle for independence, particularly under the influential leadership of Mahatma Gandhi. This noteworthy cloth is intricately linked to the historical narratives of not only India but also Bangladesh and Pakistan. Khadi, a fabric made from hand-spun cotton, represents the ideals of self-sufficiency and the unwavering

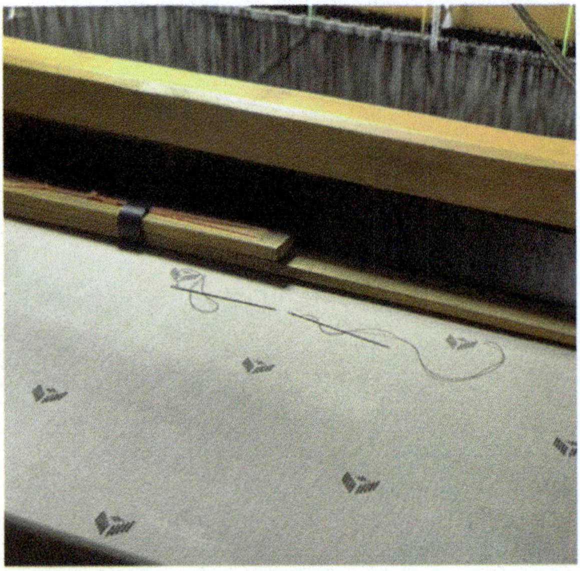

Fig. 4 Khadi fabric on khadi (by Authors)

essence of autonomy. The design language of the product has remarkable versatility, effectively connecting elements from both historical and futuristic contexts. Khadi holds significant historical value as it embodies India's unwavering pursuit of independence, transforming into an emblem of national pride and self-reliance, hence retaining its importance in the present period. Khadi, a traditional hand-spun and hand-woven fabric, is currently manufactured in many regions across Pakistan, India, and Bangladesh. These regions encompass cities such as Charsadda, Kamalia, Kasur, Jaipur, Ahmadabad, Muradnagar, and Gauripur [11] (Fig. 4).

4.1.4 Jacquard Weaving

The development of Jacquard weaving, a remarkable technological achievement credited to the French weaver Joseph Marie Jacquard, represents a significant turning point in the textile sector. The idea originated in France and played a pivotal role in the Industrial Revolution by mechanising the manufacturing process of complex woven designs. Although not as widely practised as traditional handloom or power loom weaving, Jacquard weaving established a distinct presence on the Indian subcontinent. The city of Varanasi, located in India, gained a reputation for its exceptional silk brocades, a traditional craft that flourished for several centuries, with its origins tracing back to the nineteenth century. In addition to Bangladesh and Pakistan, there are also small-scale Jacquard weaving factories located in these countries, which specialise in the production of intricately patterned fabrics for diverse applications. Jacquard weaving exemplifies the embodiment of cultural and design influences in

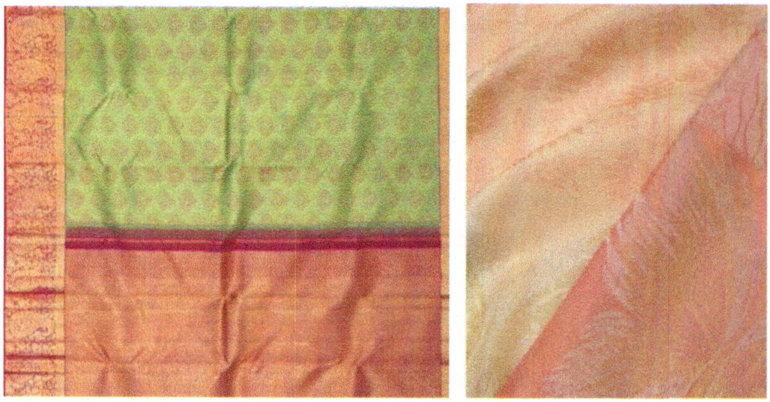

Fig. 5 Jacquard shawl and saree showcasing beautiful colors (by Authors)

the region, owing to its ability to incorporate a wide spectrum of colors and intricate patterns. The amalgamation of traditional luxury and technological innovation in this art serves to sustain the production of visually captivating textiles that hold cultural significance. These textiles not only showcase the rich cultural heritage of these countries but also contribute to their economies through exports. The combination of traditional craftsmanship and modern technology in the weaving process ensures that these fabrics continue to be highly sought after in both domestic and international markets [12] (Fig. 5).

4.1.5 Machine Weaving

Machine weaving, an integral part of the textile industry in the Indian subcontinent, has a history rooted in the colonial period. It made its debut during British colonial rule, with the British East India Company and the colonial government actively involved in the establishment of textile mills and power looms. The early years of machine weaving can be traced to the nineteenth century. These technological advancements have shaped the landscape of textile production in the region. Notably, Indian states like Tamil Nadu, Gujarat, and Maharashtra have emerged as strongholds of this industry, with Coimbatore earning the moniker "Manchester of South India" for its pivotal role in textile manufacturing. In Pakistan, industrial centers like Faisalabad and Karachi drive the textile machinery sector. Meanwhile, in Bangladesh, Dhaka is at the forefront of the textile and garment industry. Machine weaving offers unparalleled efficiency and scale, allowing the textile sector to meet the diverse demands of both local and global markets (Table 1).

Table 1 The table presents a wide range of textile weave techniques found in the countries in discussion (by Authors)

Weave	Country	Technique	Region
	Pakistan	Thari weave	Tharparkar
		Balochi weave	Balochistan
		Mangalagiri weave	Sindh
		Pochampally weave	Various regions in Pakistan
		Khas weave	Sindh
		Susi weave	Sindh
	India	Chanderi weave	Madhya Pradesh
		Paithani sarees	Paithan, Maharashtra
		Maheshwari sarees	Maheshwar, Madhya Pradesh
		Patola weave	Gujrat
		Kuchai silk weave	Jharkand
	Bangladesh	Dhakai Jamdani sarees	Dhaka (Bangladesh)
		Baluchari sarees	Murshidabad, West Bengal
		Tangail sari weave	Tangail, Dhaka
		Rajshahi silk weave	Rajshahi, Bengal
		Garo weave	Garo Hills (Mymensingh and Sylhet)

4.2 *Printing and Dyeing*

Dyeing and printing are processes employed in the conversion of raw textile fabrics into finished goods. Subcpontoinent has been known for a wide range of techniques exclusive to cities, and practitioners are taking textile design craftsmanship to the next level. The region is known for various techniques such as natural dyeing, Bandhanio, Ikat, Ajrak, Kalamkari, and more. The textile traditions of the Indian subcontinent have a rich history that spans millennia and are marketed using a diverse array of techniques. The parks have evolved and flourished over time, showcasing the cultural and artistic heritage of the region.

4.2.1 Bandhani

Tie-dye, referred to as Bandhani in India, is a traditional textile technique with historical roots in regions such as Rajasthan, Gujarat, and certain parts of Sindh in Pakistan. The origins of this custom can be traced back to the period of the Indus Valley civilization, indicating its roots and longevity. The origins of the term "Bandhani" can be traced back to its Sanskrit root "bandh," which signifies the act of binding or tying. The method entails a precise separation of fabric using fingernails to form minute bindings, which are then fastened with threads at different

Fig. 6 Bandhani dyed silk [27]

locations in order to generate elaborate designs. Bandhani artists employ natural materials like flowers and turmeric to extract pigments, each of which carries distinct cultural significance. (Bhandani and Dhingra) The primary geographical regions where bandhani-making centres are located are Gujarat, Rajasthan, Sindh, Punjab, and Tamil Nadu. In the context of Pakistan, the term "chunri" is commonly used. Various locations exhibit distinct variances in Bandhani designs and colors, with each representation showcasing exclusive patterns and cultural importance. The art in question, requires a significant amount of manual labor, is frequently passed down from one generation to the next. Women hold a crucial position in the process, as they are responsible for intricately tying patterns. This craft serves as a symbol of cultural importance, as it helps to preserve the distinct approaches to this craft seen in various regions (Fig. 6).

4.2.2 Ikat

Ikat weaving, which has evidence of existence since the Neolithic era, has existed in the subcontinent since the 19th century, with origins in India. It is a complex and lively textile culture that is widely observed in Pakistan, India, and Bangladesh. The time-honored method is widely recognised for its elaborate designs, which are achieved by dyeing the yarn prior to the weaving process. The process entails resist dyeing, wherein certain portions of the yarn are tightly tied to prevent the penetration of dye. The procedure encompasses the selection of yarn, the act of binding, and the application of dye in many colors to achieve the intended pattern. Ikat textiles exhibit vibrant hues and intricate patterns, encompassing flora, fauna, and geometric designs, wherein each locality integrates distinct cultural and historical constituents. The preservation of historical traditions and the reflection of distinct heritage in these locations are a monument to the exceptional skill of artisans [14] (Fig. 7).

Fig. 7 Ikat shawl in green and maroon (by Authors)

4.2.3 Ajrak

The art of Ajrak printing, which is predominantly observed in the regions of Sindh, Gujarat, and Rajasthan, is a time-honored technique of block printing on textiles that has been in existence for more than a millennium. The origin of Ajrak can be traced back to ancient times around 2500 BC–1500 BC. Artisans employ the technique of hand-carving wooden blocks to produce elaborate patterns, subsequently transferring them onto the cloth through the process of stamping. Ajrak textiles exhibit vibrant hues and intricate geometric motifs, frequently sourced from organic dyes. The aforementioned artistic expression continues to serve as a representation of cultural esteem and individuality in Sindh, Pakistan, as well as the Kutch region of India, hence safeguarding the longstanding artistry and customs inherent to these areas [15] (Fig. 8).

Fig. 8 Ikat samples in ranges of blue rad and off-white [27]

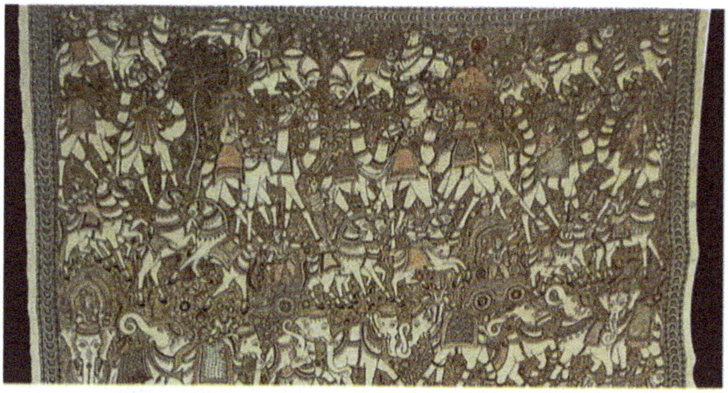

Fig. 9 Kalamkari shawl with image of elephants and fighters (by Authors)

4.2.4 Kalamkari

The craft of Kalamkari printing is a labor-intensive artistic practice that has been traditionally employed in the regions of Andhra Pradesh and Telangana for several centuries. Moreover, there is empirical data indicating the presence of this phenomenon in specific regions of Pakistan. The historical origins of this phenomenon can be traced back to a span of more than two thousand years. The procedure entails the incorporation of intricate hand-painted or hand-blocked designs with manual dyeing methods [16]. The Kalamkari designs draw influence from ancient epics, depicting diverse aspects of nature, animals, and cultural symbols, often communicating narratives or encapsulating symbolic significance. The perpetuation of this ancient tradition in India is sustained by proficient artisans, who play a crucial role in exhibiting the creative and cultural heritage of the area [17] (Fig. 9).

4.2.5 Screen Printing

Screen printing, a versatile method that originated in the twentieth century, has been used in several locations, such as India, Pakistan, and Bangladesh. The complex procedure entails the creation of a stencil or screen, commonly composed of silk or polyester, which is tightly stretched over a frame. Subsequently, the intended pattern is transmitted onto the screen, wherein meticulous measures are used to selectively block areas that are intended to stay empty of ink. After the preparation of the screen, ink or dye is subsequently applied and forced through onto the fabric, facilitating the realisation of the desired design. India is renowned for its proficient artisans who engage in the practice of screen printing throughout different locations, typically showcasing regional traditions and individual artistic expression. It serves as an essential intermediary, facilitating the integration of conventional artisanal practices with modern design and assuming a pivotal function within the textile sector.

4.2.6 Digital Printing

Digital printing is a contemporary and adaptable technique that emerged in the latter part of the twentieth century. It obviates the necessity for screens, plates, or stencils, rotary, and sublimation, offering an environmentally conscious option. The previously mentioned regions, namely India, Pakistan, and Bangladesh, have exhibited a notable inclination towards this particular occurrence, particularly within the textile and garment industries. Digital printing provides a wide range of color options and design opportunities, effectively reconciling traditional artistry with the requirements of contemporary fashion and textile industries. Prominent centres for digital textile printing are situated in places such as Surat and Ahmedabad in India, as well as Dhaka in Bangladesh, which holds a notable position within the worldwide garment industry. Moreover, the towns of Karachi and Faisalabad in Pakistan have also adopted digital textile printing techniques (Fig. 10).

4.3 Embroidery and Handwork

Embroidery is a technique where decorative stitches are added to the fabric, creating intricate and colorful designs. Different regions in India have their own styles of embroidery. For example, the delicate Chikankari of Lucknow, the vibrant Phulkari of Punjab, or the royal Zardozi work, which uses gold and silver threads for that touch of luxury [18].

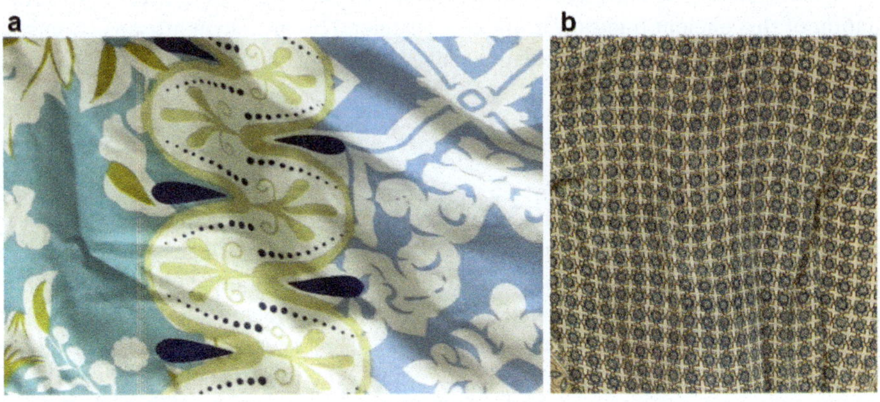

Fig. 10 **a**, **b** Screen printing shirt and digital printed bedcover (by Authors)

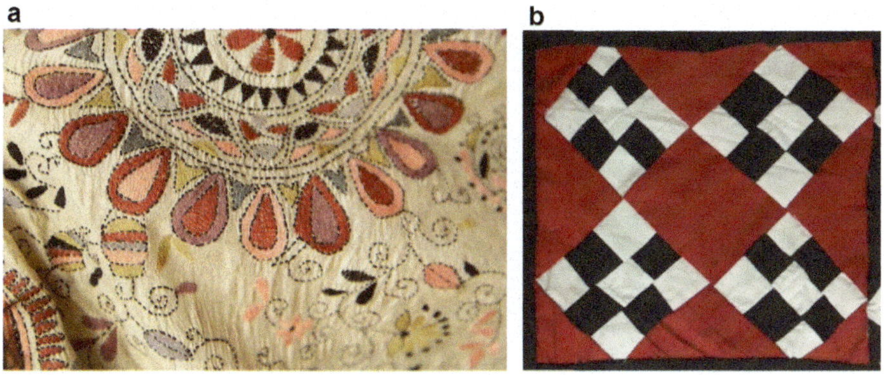

Fig. 11 **a**, **b** Nakshi Kantha shawl on left and Ralli Blanket cover on the right (by Authors)

4.3.1 Nakshi Kantha Embroidery (Ancient Tradition)

Nakshi Kantha embroidery is a traditional craft in Bangladesh, originating from Jessore and Khulna. This craft is more than just embroidery; it's a storytelling form, using stitched motifs to tell unique narratives about rural life, mythology, folklore, and nature. The artisans, mostly women, use needles and thread to weave tales with colorful stitches. Nakshi Kantha captures the essence of Bangladesh's rural lifestyle and mythical tales passed down through generations. The art form is cherished in Jessore and Khulna, spreading its influence throughout the country.

4.3.2 Rilli Work (Ancient Tradition)

Rilli is a traditional craft in India, primarily focusing on making blankets by artisans. Originating in the southern provinces of Sindh, Balochistan, Cholistan Desert, Punjab, Gujarat, and Rajasthan, Rilli embroidery is characterized by vibrant colors and intricate geometric designs. Each piece is a work of art, often symbolizing the artisan's surroundings and personal experiences. Rilli embroidery reflects the rich cultural tapestry of Sindh and the traditional skills of its artisans. Although originating in Sindh, Rilli embroidery is practiced in other parts of Pakistan, India, and Bangladesh. The craft continues to thrive, adapting to contemporary preferences while preserving its traditional charm and showcasing its enduring legacy in the Indian subcontinent [19] (Fig. 11).

4.3.3 Kasuti Embroidery (Over 2,000 Years Ago)

Kasuti embroidery, an ancient craft from the Sindh region of India, is known for its intricate geometric and angular patterns, originating over 2,000 years ago. Inspired by the region's rich flora and fauna, Kasuti designs feature both symmetric and

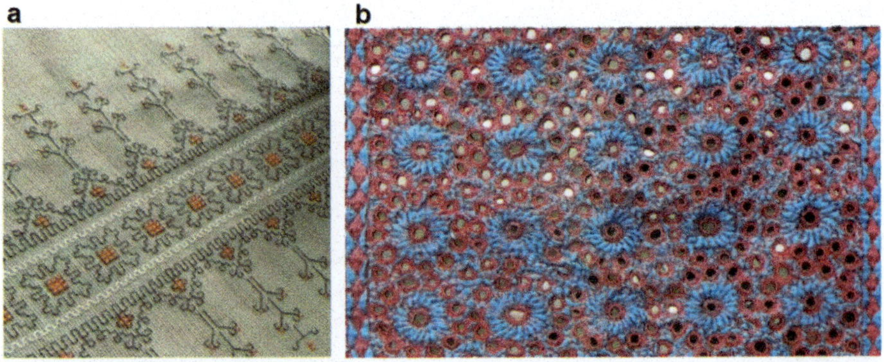

Fig. 12 **a, b** As shirt with kasuti embroidery on the left and mirror work cushion on the right (by Authors)

asymmetric motifs. The art form's geometric precision and angular aesthetics are hallmarks, exemplifying the artistic heritage of the region. Today, Kasuti embroidery is practiced in various parts of Pakistan, India, and Bangladesh, adapting it to their cultural narratives while preserving its essence. This enduring significance and adaptability of traditional embroidery techniques across the Indian subcontinent are evident [20].

4.3.4 Mirror Work (Shisha) (Over 2000 Years Ago)

Mirror work, also known as Shisha embroidery, is a centuries-old craft in India that involves embedding small mirrors into fabric to create a visually striking and reflective effect. Originating in Rajasthan and Gujarat, mirror work is skillfully integrated into textiles using colorful threads. The design language of mirror work captures and plays with light, adding sparkle and glamour to garments and accessories. This unique embroidery style has transcended borders and is still widely practiced in Rajasthan and Gujarat. In Pakistan and Bangladesh, artisans have embraced and incorporated mirror work into their repertoire, with regions like Sindh and Punjab adopting it as part of their cultural heritage. Mirror work exemplifies the enduring appeal of traditional craftsmanship in the Indian subcontinent (Fig. 12).

4.3.5 Chikankari Embroidery (Over 400 Years Ago)

Chikankari embroidery, a traditional art form in India, has a rich history dating back over 400 years. Located in Lucknow, Uttar Pradesh, it is celebrated for its use of fine muslin fabric and delicate white thread. The design language is characterized by intricate white-on-white floral patterns, creating a timeless aesthetic. One distinctive feature is the use of shadow work, where the embroidered patterns are visible on the

a b

Fig. 13 **a, b** Chicken Kari embroidery shirts on the left and sample of Zardozi work motif on the right (by Authors)

reverse side of the fabric. Chikankari is an integral part of traditional attire in India, particularly in the northern regions. In Pakistan, it is practiced in Karachi and Lahore, and is popular for its elegant appearance on clothing and accessories. In Bangladesh, it is less common but still cherished for its artistic and cultural significance.

4.3.6 Zardozi Embroidery (Over 200 Years Ago)

Zardozi embroidery, originating over 200 years ago in Uttar Pradesh, India, is an opulent and intricate style that is practiced in Pakistan and Bangladesh. This technique uses gold and silver threads, gems, and pearls to create luxurious designs with intricate floral and geometric patterns. Zardozi embroidery is a symbol of regal elegance, particularly prevalent in bridal wear and ceremonial outfits in India. In Pakistan, it is found in Karachi and Lahore, adorning traditional attire and accessories. In Bangladesh, it is less common but still cherished for its artisanal excellence, especially in high-end ethnic fashion and home decor. Zardozi embroidery continues to embody timeless luxury and exemplify the artistic mastery of the Indian subcontinent's craftsmanship (Fig. 13).

4.3.7 Phulkari Embroidery (Over 100 Years Ago)

Phulkari embroidery, a form of flowerwork with a rich history, originates from the Punjab region of India. Traditionally, it was used by women to create colorful shawls and dupattas, filled with vibrant patterns and embroidered with untwisted silk floss on a coarse cotton base. This art form is primarily seen in Punjab, Haryana, and Rajasthan, and is an essential part of the region's cultural heritage. The art has spread to Pakistan, particularly in Punjab province, where artisans continue the tradition. In Bangladesh, although not as widespread, it is still celebrated as a remarkable artistic

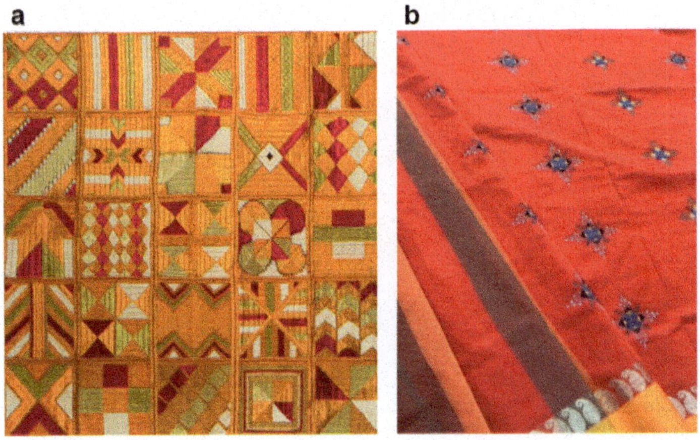

Fig. 14 **a, b** Phulkari embroidery sample on the left and Tangail Sari on the right (by Author)

expression, particularly among communities with roots in the Indian subcontinent. Phulkari embroidery reflects the cultural diversity and shared heritage of the Indian subcontinent, preserving traditional skills and fostering appreciation for the art form across borders.

4.3.8 Tangail Sari Embroidery (Over 100 Years Ago)

Tangail Sari Embroidery, a textile tradition originating in Dhaka, Bangladesh, is celebrated for its vibrant aesthetics. Originating over 100 years ago, it features intricate patterns inspired by the natural world and cultural motifs. The designs often incorporate local flora and fauna, creating a deep connection to the region's cultural heritage. The vibrant colors add energy and vibrancy to the fabric, making Tangail Sari saris a prized and popular art form in Bangladesh and neighboring regions [21] (Fig. 14).

4.3.9 Machine Embroidery and Computer Embroidery (20th and 21st Century)

Machine embroidery, a process using computerized machines to create intricate designs on fabric, has significantly impacted the textile and garment industries in India, Pakistan, and Bangladesh during the late 20th and early twenty-first centuries. Initially, semi-automatic and manually programmed machines allowed for faster and more consistent production of embroidered designs compared to traditional hand embroidery techniques. However, the real transformation came with the adoption of fully computerized machines.

a b

Fig. 15 **a**, **b** Machine embroidery shit on the left and computer embroidery shit detail on the right (by Authors)

In Surat, Gujarat, the textile industry in India saw a surge in machine embroidery workshops, allowing for more intricate designs and faster production. Pakistan's textile industry, with major centers in cities like Karachi and Faisalabad, embraced computerized embroidery machines, providing a competitive edge in the production of high-quality, intricately embroidered textiles. The availability of skilled labor and a growing number of embroidery machine manufacturers further facilitated the progress of computerized embroidery in Pakistan.

Bangladesh, known for its vibrant garment industry, also incorporated machine embroidery into its production processes, allowing manufacturers to diversify their product offerings and gain popularity both domestically and internationally. The progress of machine embroidery in these countries reflects the global trend toward automation and technological advancement in the textile industry. Combining technological innovation with traditional craftsmanship, machine embroidery has become an integral part of these countries' textile production landscape, improving productivity, and increasing their competitiveness in the global textile market (Fig. 15 and Table 2).

5 Iconic Patterns and Motifs

The utilisation of patterns and motifs for the purpose of surface embellishment in textiles has been a significant means of extracting valuable historical insights into the region for an extended period of time [22]. The mentioned factors collectively enhance the comprehension of the concepts, interests, and inclinations prevalent

Table 2 The table presents a wide range of textile embroidery and handwork techniques found in the countries in discussion "by Author"

Embroidery and handwork	Country	Technique	Region
	Pakistan	Sindhi embroidery	Sindh
		Gara embroidery	Karachi
		Kashmiri embroidery	Kashmir and parts of Punjab
		Thar embroidery	Tharparkar, Sindh
		Suzani embroidery	Karachi and parts of Punjab
	India	Banjara embroidery	Rajasthan, Telangana, and Andhra Pradesh
		Toda embroidery	Nilgiri Hills, Tamil Nadu
		Mukesh work	Gujarat
		Sujuni embroidery	Bihar
		Lacework (Chikan Work)	West Bengal and Lucknow
	Bangladesh	Aari embroidery	Dhaka (Bangladesh)
		Baluchari sari embroidery	Murshidabad, West Bengal
		Mizoram hand embroidery	Sylhet and parts of Bangladesh
		Chikan work	Dhaka and other urban centers

within the region, which accommodates a diverse array of faiths and civilizations. The utilisation of motifs and patterns in art is indicative of conceptual and visual considerations, as well as the cultural context that influences both the creator and the wearer. A few significant features are discussed below [23].

5.1 Buta (Paisely or Ambi)

The design motif known as paisley, a textile pattern characterised by colorful, curved abstract figures, is taken from the ancient Aryan boteh (botteh) motif. 'Boteh' is a Persian word meaning bush, shrub, cluster of leaves, or a small, dense forest of small trees or bushes. Due to huge shawl production in Paisley, Scotland, the pattern was given the name paisley. Paisley, an important motif widely used in Indian textiles and embroidery, is a teardrop-shaped motif that resembles a mango, so it is called mangai motif or Ambi, which means mango in Punjab. It is also known as a carrey design, named after the Urdu word for mango seed. Whether adorning Kashmiri shawls in India, gracing the textiles of Pakistani Punjab and Sindh, or weaving stories through nakshi kantha embroidery in Bangladesh, paisley is a testament to its enduring charm

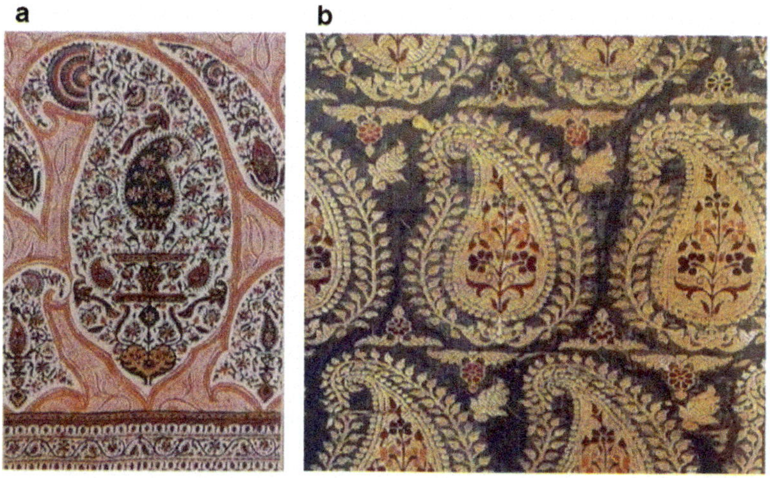

Fig. 16 **a**, **b** Printed buta on the left and Banarasi woven buta sample on the right (by Authors)

and cultural significance in the Indian subcontinent's rich tapestry of crafts. Symbolic motifs in traditional Indian textiles and embroidery [24] (Fig. 16).

5.2 *Animal Motifs*

Animal motifs adorn the textiles of India, Pakistan, and Bangladesh, serving as representations of the local fauna and cultural heritage. The previously mentioned motifs of elephants in India, peacocks in Pakistan, and an assortment of fauna in Bangladesh encapsulate the magnificence of nature and tradition. By portraying the local environment, the artwork strengthens the connection between culture and nature. Elephants symbolise wisdom, fertility, strength, and prosperity, in addition to nobility. Peacocks represent protection from malevolence and rebirth. The motif of a fish represents chastity and procreation. These motifs represent the eternal connection that exists between nature and culture. Animal motifs are frequently incorporated into embroidery techniques, including Kalamkari, Ajrak, and Sujni, among others (Fig. 17).

5.3 *Floral Patterns*

Floral motifs are a common feature in textile crafts across the Indian subcontinent, including India, Pakistan, and Bangladesh. These motifs represent various emotions, beliefs, and aspects of life and are associated with deities, rituals, and celebrations in

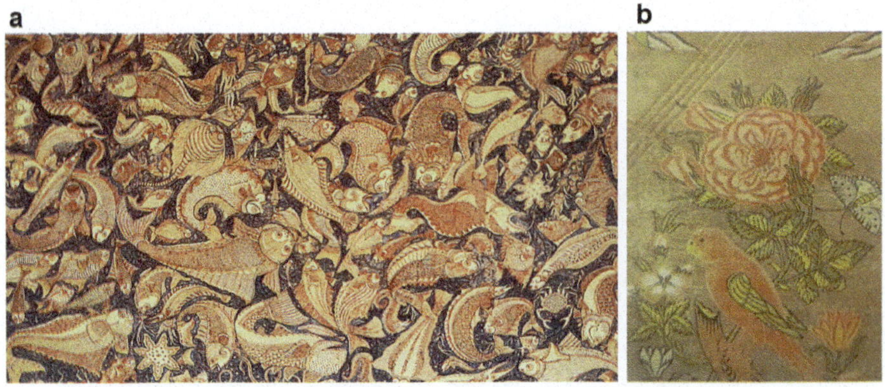

Fig. 17 **a, b** Two sample images showcasing Animal Motifs (by Authors)

Indian mythology and religious traditions. Examples of these textile crafts include Chikankari embroidery, a delicate white-on-white pattern from Lucknow, India; Phulkari embroidery, a vibrant pattern from the Punjab region and parts of Pakistan; Nakshi Kantha embroidery from Bangladesh, often depicting nature, rural life, and mythology; block printing techniques from India, such as those in Rajasthan, often incorporate floral patterns in their designs; Tangail Sari embroidery from Bangladesh, featuring intricate floral motifs inspired by the natural world and cultural traditions; and Sanganeri printing from Rajasthan, India, known for its detailed floral designs [25] (Fig. 18).

Fig. 18 A Kalamkari textile surface with floral pattern (by Authors)

Fig. 19 Khadi Woven geometric pattern surface (by Author)

5.4 *Geometric Patterns*

Geometric motifs are prevalent in various textile crafts across the Indian subcontinent, including India, Pakistan, and Bangladesh. Ikat weaving, Kantha embroidery, Jamdani, Phulkari embroidery, Ajrak printing, block printing, and Tangail Sari embroidery are some examples of these crafts. Ikat textiles from Gujarat and Odisha showcase intricate geometric patterns using tie-dye and resist-dye techniques. Kantha embroidery from Bangladesh incorporates geometric motifs in its stitching patterns, ranging from simple lines to complex designs. Phulkari embroidery from the Punjab region includes geometric motifs in its vibrant patterns. Ajrak printing from Sindh, Pakistan, features intricate squares and diamonds. Block printing techniques from India, such as those in Rajasthan, also incorporate geometric patterns (Fig. 19).

5.5 *Tree of Life*

The Tree of Life is a prominent symbol in various artistic disciplines, including textile design, handicrafts, and fine arts. Originating in India, it represents a vertically oriented tree with branches that extend outward. Drawing inspiration from Hindu mythology, Islamic tradition, and the Bodhi tree, it has been used as a biomorphic motif since the mid-sixteenth century. The symbol is used in various crafts, including textiles, chintz, silks, and printed cloth, using techniques like batik, block printing, mordant painting, handweaving, and embroidery [26] (Fig. 20).

Fig. 20 A kalamkari textile sample portraying the tree of life (by Authors)

6 Discussion

The growth of textile design in India, Pakistan, and Bangladesh exemplifies the intricate interplay of culture, craftsmanship, and artistic expression. Every nation has cultivated its own unique styles and methodologies, resulting in the emergence of emblematic designs and symbols that embody indigenous plants, animals, and customs. The themes of paisley, animal symbols, flower patterns, and geometric designs are intricately integrated into the fabric of everyday existence, carrying profound cultural importance.

Also, textile design shows how nature and culture have always been connected; different patterns represent different parts of life, beliefs, and rituals. The skilled craftsmen who engage in these artistic pursuits perpetuate age-old traditions that have been transmitted over successive lineages, enhancing the cultural fabric of their individual localities.

When you look more closely at the complicated steps needed for weaving, dying, printing, and embroidery, you can see that they are real examples of how cultural traditions can last for a long time and how craftsmanship can change along with the fast-paced changes in today's world.

Textile design in the Indian subcontinent encompasses more than just an economic sector; it serves as a story that reflects the region's rich heritage, cultural identity, and artistic ingenuity.

References

1. Gillow J, Barnard N (1993) Traditional Indian textiles. Thames & Hudson Ltd.
2. Riello G, Roy T (n.d.) How India clothed the world: the world of South Asian Textiles, 1500–1850. 2009: Brill
3. Wilson J (2001) Handbook of textile design. Woodheld Publishing Limited, Cambridge
4. Studd R (2002) The textile design process. Des J 35–49
5. Hann M (2021) Textile design products and processes. CRC Press, London
6. Dhamija J (2014). Sacred textiles of India. The Marg Foundation.
7. Crill R (2015) The fabric of India. V&A
8. Agarwal Y (2008) Silk brocades. Roli Books Pvt Ltd
9. Gandham V (2022) The legacy of banarasi motifs and patterns. From Singhania's. https://singhanias.in/blogs/singhanias-saree-journal/the-legacy-of-banarasi-motifs-and-patterns
10. Weaving of Bangladesh (n.d.) From Asia Inch. https://asiainch.org/craft/textiles-of-bangladesh/
11. *Khadi* (2023) From the craft atlas. https://craftatlas.co/crafts/khadi
12. Agrawal Y (2008) Silk brocades. Roli Books Pvt Ltd
13. Bhandani V, Dhingra S (n.d.) Textiles and crafts of India. Prakash Book Depot, New Delhi
14. Books I (2011) World textiles: a sourcebook. Interlink Pub Group Inc
15. Legrand C (2012) Orissa, Rajasthan, Gujarat: the colours of India. In: Legrand C (ed) Textiles a world tour. Thames and Hudson
16. Sing M, Chisti RK, Jain R (2011) Traditions and beyond: handcrafted Indian textiles. Roli Books, New Delhi
17. Sengupta R (2016) Kalamkari: textile tradition of Andhra Pradesh. From sahapedia. https://www.sahapedia.org/kalamkari-introduction
18. Waller D (2012) World textiles. British Museum Press
19. Mubashra A (2015) Ralli: a legendary craft nurtured by the marginalized. THAAP J
20. Create D (2016) Kasuti embroidery from Karnataka. https://medium.com/direct-create/kasuti-embroidery-from-karnataka-9edd4a50da07
21. Tangail Saree (n.d.) From government of West Bengal. https://www.westbengalhandloom.org/htm/h_tangail.html#:~:text=Tangail%20Saree%20are%20famous%20with,Districts%20in%20the%20undivided%20Bengal.
22. Ali A, Malik A, Aamir N (2017) The ideological basis of design motifs in the architecture of subcontinent: the converging stories of bells and tassels. Pak Soc Sci Rev
23. Fotheringham A (2019) The Indian textile sourcebook: patterns and techniques (Victoria and Albert Museum). Thames and Hudson Ltd.
24. Editors E (2022) Buta From Mapacademy. https://mapacademy.io/article/buta/
25. Veenu KC, Sharma RB (2016) Symbolic motifs in traditional indian textiles and embroideries. Int J Res Econ Soc Sci
26. Motifs Speak: a Vocabulary of Indian Textile Motifs (2022) From Selvedge. https://www.selvedge.org/blogs/selvedge/motifs-speak-a-vocabulary-of-indian-textile-motifs
27. Neumann K (n.d.) The Bandhani collection. From house of wandering silk. https://www.wanderingsilk.org/ajrakh-collection

Textile Technologies and Amplifications

Suniya Tariq and Fatima Iftikhar

Abstract The textile industry is a significant contributor to economic growth in South Asian nations, particularly Pakistan, India, and Bangladesh. Their roots date back to the late eighteenth century during the industrial revolution and have a rich history, with artifacts from the Indus Valley civilization showcasing elaborate weaving methods and artisanship. The British colonial government in the nineteenth century introduced technological innovations like the Mule Spinner, laying the groundwork for the industry's transformation. In the twenty-first century, the textile sector has seen unprecedented technological advancements. The chapter examines the technological advancements that occurred in these nations from the 19th to the twenty-first centuries, with a particular focus on their contribution to the development of sophisticated textile solutions that cater to the demands of diverse applications, industries, and the global community. This encompasses their importance in mitigating environmental issues linked to conventional synthetic fibers, thereby demonstrating the area's dedication to sustainable methodologies.

1 Background and Significance (Technological Advances in Textile Industry)

The main pillar of the economic development of South Asian countries is the textile industry. The textile industry of regions such as Pakistan, India, and Bangladesh has been prospering over the years with new advancements in technological aspects of the textile sector. The development of the textile industry in this area started in the late eighteenth century, during the era of the industrial revolution [1].

S. Tariq · F. Iftikhar
Pakistan Institute of Fashion and Design, Lahore, Punjab, Pakistan
e-mail: s.tariq@deakin.edu.au

S. Tariq · F. Iftikhar (✉)
Deakin University, Melbourne, VIC, Australia
e-mail: f.iftikhar@deakin.edu.au

H. Jamshaid and A. Dad (eds.), *Creative Textile Industry*, SDGs and Textiles,
https://doi.org/10.1007/978-981-97-3802-1_5

Many textiles have been found on the subcontinent since prehistoric times. The first proof of the presidents of textiles was found in the Indus Valley civilization (3300 BCE–1300 BCE) in India. The archaeologist discovers the existence of various types of weaving techniques made with intricacy using distinct fibres. Illustrating the proficiency and expertise in artisanship in handwork fabrics in the region [2]. Illustrating the narrative of history in the development of textiles in Pakistan has also interweaved its evidence in prehistoric times and can be seen in the influences of Persian and Mughal art, architecture, and clothing with skilled craft of weaving techniques, natural dying with vibrant colours, and intricacy in design and patterns showcasing the rich heritage of the region. Likewise, the history of Bangladesh in textiles is also deeply inherited and rooted in the past. The evidence of traditional handloom weaving has been a significant part of that area, renowned for its sheer Muslin cloth that was made with delicacy with extremely fine hand spun yarns and is often known as "woven air" [3].

As we transition through the time between the emergence of British rule and the colonial era, the textile industry of the subcontinent experienced a major change. The emergence of East India Company witnessed the economic potential of the textile trade in the region, exploited the local resources, and declined the traditional handloom industry by powering up the industrial looms. In the 1760s, the Industrial Revolution marked a turning point in the textile industry by revolutionising high-speed looms such as the spinning Jenny, which was invented by James Hargreaves and provided multiple spinning frames that allowed textile production to increase its efficiency [4].

A progressive transformation was seen in the development of technology in the region of the subcontinent, with each step moving forward towards increasing bulk production and extravagant quality. The following table shows the chronological order of inventions made over the years in the textile industry (Table 1).

In the early 1800s, the subcontinent saw the introduction of power looms, which were invented by Edmund Cartwright to enhance the weaving process. This development of the power loom also accelerated and provided better efficiency in the textile production of the subcontinent, and the manual labour shifted to the mechanical systems of the loom.

As we move forward in the twentieth century, the mid-1900s also witnessed the adoption of synthetic fibres replacing the natural fibres, like nylon, which was invented in 1935, and polyester, which was developed in the early 1940s, which became the major ingredient of textile manufacturing because this synthetic material offered a range of new properties such as durability and versatility, revolutionising the industry by contributing to the mass production of textiles. The latter half of the twentieth century started a new era of technological advancement in the textile sector, innovating new features of integration of computer-aided designs (CAD) that allow the development of designs and industrial processes to be more precise [6].

In the latest twenty-first century, digitalization processes have been increasingly marking up their platform in the innovations of smart textiles and technical textiles industry. With the emergence of functional designs like embedment of sensor fabrics, sensing capabilities in textiles, eco-friendly sustainable approaches, waste and water

Table 1 Technology evolution in sub-continent (India), from 2nd—late seventeenth century (by authors) Adapted from [5]

Timeline	Evolution in technology
2nd–6th Century	The carder's bow was introduced
11th Century	Earliest reference to a weaving site inscription
1184	Evidence of the operation of the vertical loom
11th Century	The application of resins to confine colours to patterns and of mordants to take colours, very common during the century
12th Century	Clear reference to the tie and dye technique
12th Century	Probable that block printing existed
1173 and 1223	Dyeing gradually an independent profession-dates record tax imposed on dyers
12th Century	On the basis of inscriptional evidence, the patterned or draw loom can be traced back "positively", introduced by Muslims from Persia
13th Century	Reference to brokers in the textile trade
1257	The spindle wheel was in use in Persia
1270	The spinning wheel in its simplest form is illustrated in China in a manner indicating its general use
1301–2	Reference to the hand-spindle
1313	Multi-spindle wheels illustrated in China
13–14th Century	Two important instruments for cleaning cotton, were the wooden worm-worked roller (charkha) and the bow scotch (Kaman). They had come into use by the centuries
14th Century	The spinning wheel which was so crucial to the increase in yarn production came with the Turks
14th Century	Calico printing
14th Century	Evidence of spinning wheels powered by water
14th Century (end)	Earliest tax on looms (Tamil region)
1426	Evidence of concentration of weavers around temples and assignment of streets
1480	The U-shaped flyer rotating around the spindle evident in Europe
15th Century	Evidence on emergence of weavers-merchants and operation of several looms at one site
16th Century	First mention of tax on carders

(continued)

Table 1 (continued)

Timeline	Evolution in technology
15–16th Century	The invention of the draw loom appears to be an indigenous achievement
1538	Inscription from Tirupati shows specialization in weaving with the use of a draw loom—probably imported by Muslims
17th Century	Looms identical with Europeans since 13thCentury depicted
17th Century	Weavers' residential shift from temple premises to coastal villages and company settlements
Late 17th Century	Alterations to looms in Madras and Fort St David to fit the demands of European market

conservation and preservation processes in dyeing, and efficiency of energy limitations in machinery, a notable change is coming forward in the textile sector of South Asian countries [7].

Assuming now, as we look forward, the textile industry of the South Asian countries is at the edge of new industry 4.0. Artificial intelligence (AI) is now redefining production processes with enhanced, optimised, and customised supply chains in the textile industry. The new era of industry 4.0 provides an overarching structure for innovating and exploring methods to advance textile production in these regions. The adaptability of Industry 4.0 is certain and is growing day by day in the textile industry of these regions [8].

2 Adoption of Technology in the Textile Industry

2.1 A Technology Advancement: Fibre to Yarn

In the global textile industry, the transformation period was introduced in the late eighteenth century, when spinning machines were introduced. In the South Asia region, the rich and extensive history of textile artisanship evolved into the revolution of spinning technology that took place through the transition of the Industrial Revolution [9].

Spinning Jenny (1764): Invented by James Hargreaves, it allowed an increase in yarn production efficiency due to the development of multiple spindles working together simultaneously [10] (Fig. 1).

Water frame (1769): Invented by Richard Arkwright, which incorporated spinning for yarn production through the power of water, is considered a new innovation for making improved stronger and thinner thread [9] (Fig. 2).

Fig. 1 Spinning Jenny (1764) (by authors)

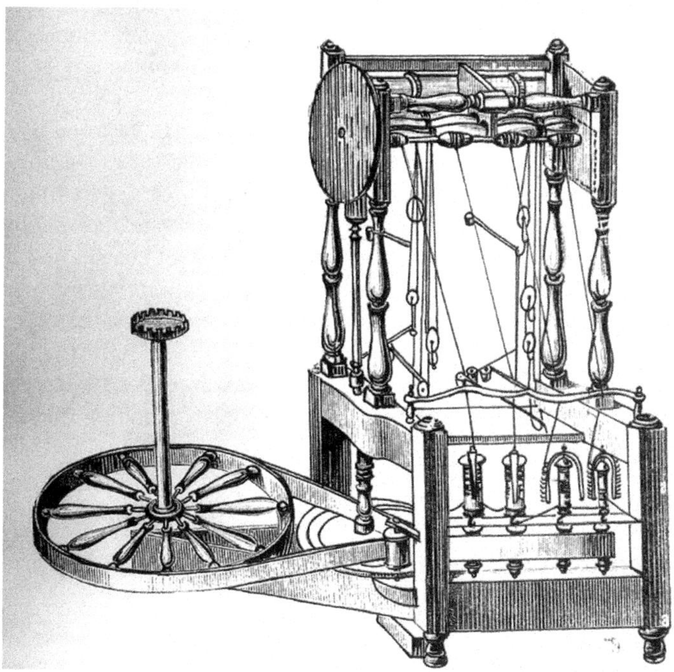

Fig. 2 Water Frame (1769) (by authors)

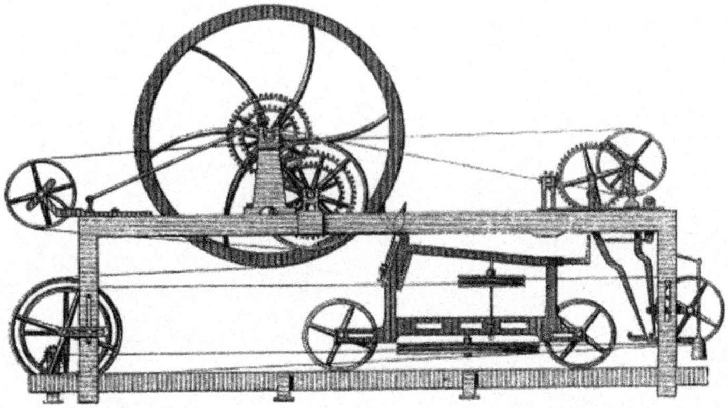

Fig. 3 Mule Spinner (1779) (by authors)

Mule spinner (1779): Invented by Samuel Crompton, it was a combination of new features and had the characteristics of both Jenny and a water frame, allowing the development of delicate, finer yarns with durable and strong properties [11] (Fig. 3).

The era of the nineteenth century marked the British colonial rule that brought the advancement of technology in spinning to the subcontinent to enhance the development of yarn production and increase the efficiency of the machines.

In the twenty-first century, spinning machines have evolved and adapted new advanced technology to produce and develop varieties in yarn count and fineness. Each machine has its own mechanism and differs in its spinning process, producing high quality.

Compact spinning machines: used in the development of yarn quality to improve the durability and strength, along with the evenness of the yarn.

Open-End Spinning Machines: Provide effective efficiency in high production of the yarn along with the development of thicker and coarser counts in the yarn.

Ring spinning machines: are used in the production of coarser and finer yarns, with the automatic software reducing manual labour and giving mass production. It is usually used in the making of cellulosic fibres like cotton, bamboo, etc. [12] (Fig. 4).

Rotor spinning machines: are used in the development of large-scale yarn production with a very high speed capacity. It produces a homogeneous yarn with less hairiness [13] (Fig. 5).

2.2 *Innovation in the Development of Fibre and Yarn*

The development of fibre and yarn in the South Asia region has seen a diverse transformation in the production of yarn with enhanced technology. The region is rich in producing natural yarns like cotton, jute, bamboo, and blends. India is a step

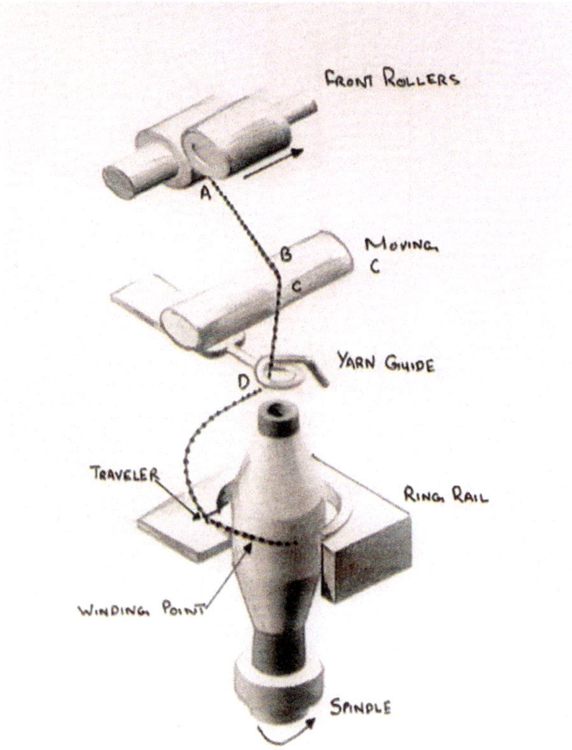

Fig. 4 A modified ring spinning system (by authors)

ahead in the manufacturing of synthetic yarns like nylon and polyester. The region is also practicing sustainable approaches for production processes to make a mark globally.

2.2.1 Natural Fibres

South Asian countries have a landmark in agriculture and cultivation of cotton, jute, flax, and hemp. Along with the large-scale development of wool and silk, Cotton is considered a key component of South Asia and the economic development of the textile industry. India and Pakistan are considered the world's largest developers and producers of cotton fibre, especially with its unique and mature qualities. Jute fibre is also produced at mass production, especially in Bangladesh, which contributes 30% of the world's production of hessian and burlap cloth. Other than this, 65% of kenaf and jute fibre are produced in India in the regions of Bihar, Assam, Orissa, and Andhra Pradesh. The lush, fertile soil and humid, warm climate make the germination rate and active growth of cotton very high in these regions. Similarly, the production of silk yarn is considered the lustrous, luminous, and radiant polished yarn of this area,

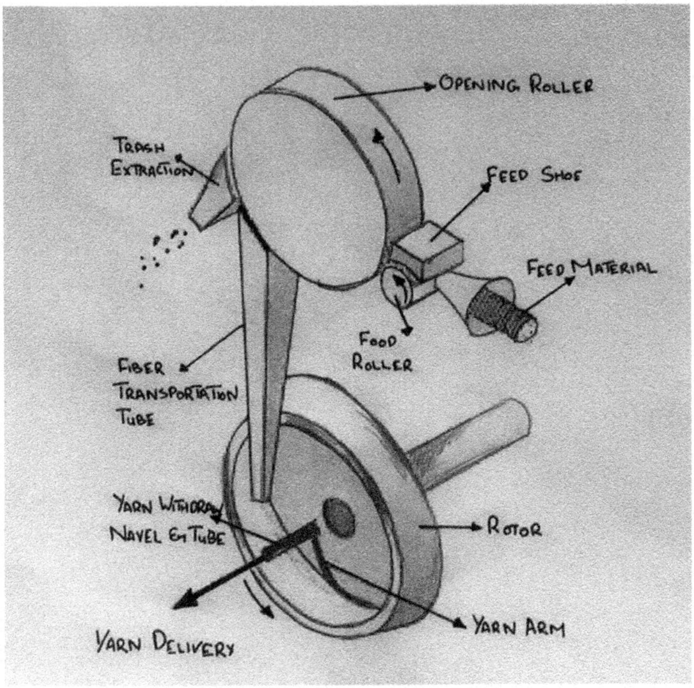

Fig. 5 A modified rotor spinning system (by authors)

with the majority of its development occurring in the states of Karnataka and West Bengal in India. Pakistan also produces silk yarn and has home-based industries near the city of Changa Manga. India also has a large production of coir fibre, mainly produced in the coastal regions of Kerala. Around 71% of coir is produced every year in India. India also engages in the cultivation of flax fibre and produces 9% of flax yarn worldwide. Natural fibre heritage is outgrown in South Asian countries, reflecting the traditional artisanship of the area with the evolution and adaptation of new technology [14] (Fig. 6).

2.2.2 Synthetic Fibres

The South Asian region, especially India, has shown a diverse development of synthetic fibres and has a vast industry to produce the yarn in mass production. The production of synthetic yarn and filaments is the strength of the industry in India, making it the fifth largest producer in the world. The textile sector develops the raw materials for synthetic filament and spun yarns. The Indian textile industry offers a variety of distinct synthetic and man-made filament yarns, including polyester, nylon, acrylic, and viscose. All the filaments have complex structures with multi-fibre properties and characteristics for a variety of products and textile applications

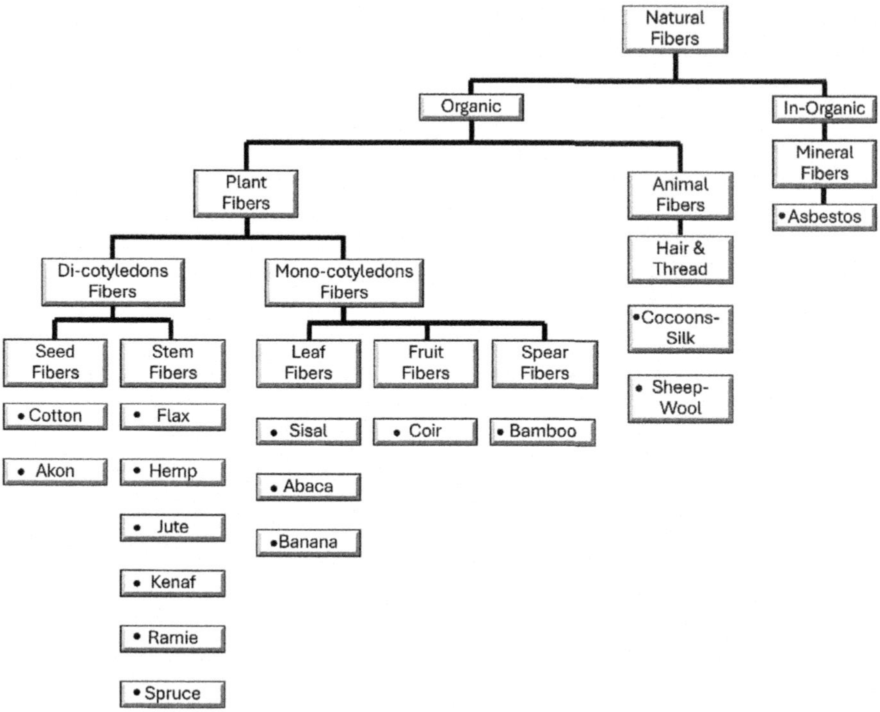

Fig. 6 Some major natural fibres found in South Asian region (by authors). Adapted from [15]

[16]. Besides this, India is also a producer of rubber used in the development of Lycra and latex yarn and fabric. India has an 8% share of the production of rubber worldwide. Other than this, Pakistan's textile industry is also producing blended yarns with man-made fibres through their spinning facilities, especially with polyester.

2.2.3 Sustainable Development of Natural Fibres with Reinforced Composites (NRFCs)

In the textile and manufacturing industries, natural fibre-reinforced composites, or NRFCs, emerged as an innovative development in which natural fibres in the form of composite materials are used. These natural reinforced composites can easily be replaced by composites made from synthetic fibres such as carbon, glass, polyester, and nylon. They are considered sustainable alternatives due to advantages such as emission of a lower carbon footprint in the environment during production and are considered an emerging source for making renewable materials [17].

NRFC's play a pivotal role in the development of ecologically and biodegradably friendly environments with the use of natural fibres obtained from renewable resources. As the dependency on glass fibre and carbon fibre has increased day by day due to its vast use in industries like aerospace, marine automotive, and construction, the use of these materials in extensive need has also increased water waste and other hazardous concerns. To address these issues, NRFCs have been a contributing achievement in the development of sustainable textiles in the form of composites that can take the place of materials generated from synthetic fibres [18].

Natural fibres like flax have superior properties and characteristics in many categories and can address smart issues such as acidification, global warming, and ozone layer depletion in comparison to glass fibre. Similarly, natural fibres such as jute, banana, and coir sisal have a very rich and high production in South Asian countries like India, Bangladesh, and Pakistan, making them the power station to utilise natural fibres for benefits like composite material with its applications in various industries such as construction, aerospace, automobile, oil, gas, telecommunication, electronics, sports, biomedical, security, and military [18] (Table 2).

The NRFCs have been a significant attraction to automobile companies for the manufacturing of a variety of components. Due to the lightweight, durability, high strength, stiffness, resistance to corrosion, and sustainable production properties of NRFCs, famous automobile manufacturing companies like Mercedes, Toyota, Ford, Audi, Mitsubishi, Volvo, Peugeot, and Fiat have shifted their production towards natural composites and are progressively expanding their business to the South Asia region due to their expertise [18].

The development of materials like NRFC's in the South Asian region has importance for the sustainable and eco-friendly approach that has been practiced in the textile industry's manufacturing processes. This application has a biodegradable impact on the environment and is a key approach to achieving a circular economy on the aspects of reduce, reuse, and recycle. This shows the combination of traditional natural fibres with respect to modern innovations in the making of NRFCs, showing technological and textile industrial progress in South Asian countries [18, 19].

Table 2 Natural fibres and their industrial applications (by Authors)

Fibre	Industrial applications
Hemp	Textiles, Geotextiles, Paper and Packaging, Electrical, Furniture, Cordage, Construction Concrete Items, 3D Printing Filaments, Manufacturing Pipes, Auto Parts and Oil Absorbent Materials
Jute	Building panels, Door Frames, Chip Boards, Geotextiles, Door Shutters, Packaging, Transport and Roofing Sheets
Flax	Tennis Racket, Bicycle Frames, Snow Boarding, Panels, Doors, Laptop Cases, Printed Circuit Boards and Furniture
Bamboo	Building and Construction Items, Baskets
Banana	Furniture, Building and Construction Items
Kenaf	Mobile Cases, Installation Materials, Animal Beddings and Packaging Materials
Cotton	Textiles, Cordage, Furniture Upholstery and Goods

2.3 *Technology Innovation in the South Asian Textile Industry*

In South Asia, the handcrafted technique of weaving, which is centuries old and connected to its roots of artisanship, has also experienced a dynamic change in the evolution of technology and new machines to re-design the process of fabric for the future. This expertise enhanced the development and production quickly. The significant change in transforming hand looms into technology in machines took place in the late 18th and early nineteenth centuries, when power looms were introduced in South Asian countries [20].

Power Loom (1785): Invented by Edmund Cartwright, it was a mechanised and enhanced version of hand looms, providing a vast edge in production efficiency [21].

Jacquard Loom (1801): Invented by Joseph Marie, Loom was a software programme that allowed the development of delicate patterns, magnifying the probability of design varieties in woven fabrics [22].

In the mid-twentieth century, more innovative looms were introduced in the textile market of South Asian countries with advanced technological expertise.

Projectile Loom (1927) and Air-Jet Looms (1952): Use in the development of indicated design patterns with increased production efficiency of the fabric [23] (Fig. 7).

The latest technology has now evolved in the twenty-first century in the textile industry and fabric development in South Asian countries.

Advance Projectile Looms: These looms have automatic control panels and systems that allow the development of intricate and precise readings in fabric.

Rapier Looms: Use in the production of upholstery fabrics with various weft inputs like fancy yarns and coarser yarns [24] (Fig. 8).

Digital Air-Jet Looms: Automation in the development of woven fabrics with various design patterns, intricate fabric checks, and quality.

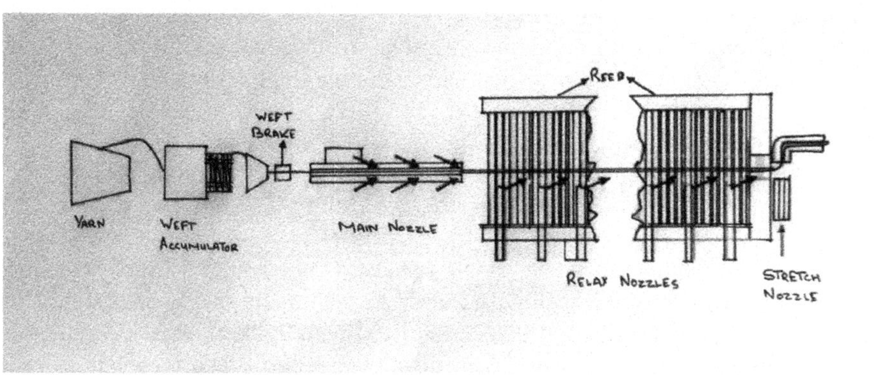

Fig. 7 Air jet loom: weft insertion mechanism (by authors)

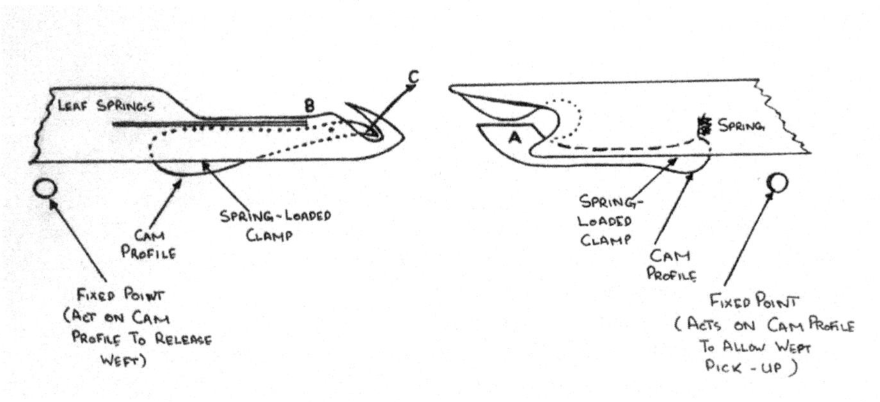

Fig. 8 Rapier loom: weft insertion mechanism (by Authors)

The hand knitting techniques used with the help of knitting needles and frames has also been evolved over the years especially in South Asian countries reflecting a traditional artisanship of the textile industry manufacturing processes. The machinery of Knitting evolved in the late nineteenth century and mid-twentieth century.

Circular Knitting (1874): The introduction of circular knitting machines allows the development of seamless and ceaseless fabrications [25]

Warp Knitting (1939): The invention of warp knitting machines allowed bulk production with diverse transformations in techniques and design developments.

In the early twenty-first century, machines become more advanced, and digitalization processes are more incorporated.

Flat Knitting: Computerised and automatic software are introduced for high-end efficiency, better quality, and various design developments [26] (Fig. 9).

Seamless Knitting: The machines helped in the development of new technology that eliminated the use of stitching and produced a proper finished garment.

In South Asian countries, many diverse transformational changes in the textile industry manufacturing processes took place that advanced the production and design development of the fabric, giving the region a cutting-edge economic benefit.

2.4 Advancement in South Asian Fabric Development

The South Asian countries have not only opted for the advanced use of machinery but have also explored new techniques in the development of various fabrics like Terry fabric, Denim fabric, and Fleece fabric. Countries like Pakistan, India, and Bangladesh have prospered and revealed outstanding progress in diverse fabric development. They have played a very important role in the manufacturing of certain fabrics, with denim being the centre of attention due to its versatility and high-end durability. In addition, Terry Fabrics, another significant fabric being produced, is

Fig. 9 Flat bed knitting machine (by Authors)

finding its use in the high-end market. These expertise and changes proved that the South Asian textile industry is not only evolving but also making efforts to cater to consumers' needs and making steady progress in the global market [27].

2.4.1 Denim Fabric

The denim manufacturing industry, especially in Pakistan, is a major exporter all around the globe, as it is a renowned denim manufacturer offering a diverse range of functional and aesthetical products used in various applications with the incorporation of sustainable practices. India also has a huge denim industry with notable practices in techniques and production. The overall South Asian denim industry is setting global trends in new techniques, focusing on quality, comfort, durability, and environmental benefits [28]. Denim manufacturing has an extensive process for development, as follows:

1. **Yarn Development**: The initial stage is the spinning of high-end cotton fibres that are opened, blended (if required), and then carded to remove the impurities to make a sliver, which then undergoes the process of spinning (ring or open end), producing a yarn for the denim formation. The cotton used in this process is of mature quality to ensure optimum strength.
2. **Warping and Dyeing**: The next phase is warping, in which the longitudinal yarns are set in the form of ropes, known as the Ball Warping Process. These ropes are involved in the dyeing process. Natural indigo dyes are used to achieve rich and variant blue colour shades in the manufacturing of denim. The dyeing processes

Fig. 10 Denim fabric and indigo shades (by Authors)

are different depending on the manufacturer, like continuous dyeing or slasher dyeing.

3. **Sizing:** After dyeing, the warping yarn is sized to enhance its strength and durability. This warp yarn is then wound on the beam, ready for the weaving process.
4. **Weaving:** The blue-dyed warp is interlaced with white weft yarn to create denim fabric. The weave used here is usually 2/1 Z twill to achieve the diagonal effect in weaving along with strength [29].
5. **Finishing:** The fabric then undergoes different finishing processes, like singeing, to remove the impurities and enhance the texture and appearance of the fabric. Washing and de-sizing are done to achieve the desired comfort and feel of the denim fabric (Fig. 10).

2.4.2 Sustainable Advancement: Evolution in Denim Fabric

In Pakistan the production of denim fabric has faced many environmental hazardous challenges like water waste and chemical management being the top priority. So the Industry has evolved its manufacturing to a sustainable practice to alleviate atmospheric impacts. Major sustainable applications that are utilised by Pakistan Denim Industry are:

1. **Energy Efficiency**: For the machinery operational process the manufacturers are investing in major energy efficient technologies like heat recovery systems and geothermal renewable sources like wind turbines, solar panels and fossil fuels.
2. **Water Conservation**: As the production of denim use rigorous water, techniques like closed loop systems, recycling of water, and low water dyeing are adopted for the preventing water waste and pollution.
3. **Chemical Management**: The Dyeing process of denim uses extensive dyes to achieve colours like red and greens other than blues with the help of sulphur etc.

Table 3 Pakistan textile and sustainable development in Denim industry [30]

Sr. No	Industry	Sustainable development
1	Naveena Denim	100% waste water treatment plant, 100% chemical management and use of solar panels for energy efficiency practice
2	Soorty Enterprises	40% reduction in water wastage. Chemical and energy consumption recycling sustainable applications
3	Artistic Milliners	Incorporated sustainable technologies like laser printing and ozone washing foe reduction of water and chemical pollution

Many industries are now using eco-friendly ways for the disposal of synthetic dyes to attain an environmental friendly environment with low health risks.

4. **Fair Labour Practice**: To raise voice for the labour rights, the industry is fulfilling the requirement of healthy working environment, adequate income and equality at their work place.
5. **Product Traceability**: Due to the involvement of complex supply chains, the challenges to trace the origin of the material difficult. The manufacturers have started implemented the RIFD tags to track the material easily keeping in mind the social standards.

It is only a matter of time that these challenges are being overcome by the denim manufacturers in Pakistan, involving all the required sustainable practices in the production of Denim (Table 3).

2.4.3 Terry Fabric

3D fabric development is also one of the captivating innovations in textile surface development that have been practiced in South Asian countries. Terry Fabric is one of the unique 3D fabrics that have been developed in these countries. Pakistan and India especially emerge as the best Terry fabric manufacturers due to its high quality, durable fabric, high absorbency, and amazing fabric texture. The innovation in the development of 3D surfaces not only enriches the physical appeal but also provides a great advantage in the functional characteristics of the fabric. They are a global market and an exporter of Terry fabrics [31].

Following are the key properties of Terry Fabrics:

1. High Water Absorbency
2. High water retention
3. Quick Drying
4. Soft and comfortable
5. High-quality surface texture.

The Terry Fabric manufacturing process for the development of 3D surfaces is as follows:

1. **Warping:** Two different beams are used in the making of Terry fabrics. One is known as the ground warp, and the other is called the pile warp.
    - **Ground Warp**: This beam makes the base of the fabric, and it is highly tensioned to avoid entanglements and make the fabric durable.
    - **Pile Warp**: This beam makes the upper surface of the fabric, and it is lightly tensioned to make loops or piles for the fabric.
2. **Sizing:** The ground warp is sized only with a safety coating to enhance the strength of the yarn and make it more resistant to friction during the weaving process.
3. **Dual Beam Mechanism**: Both the warps are wound on separate beams and have different mechanisms from each other. They are then connected to the weaving machine.
4. **Weaving:** Weft yarn is inserted in such a way that it interweaves all the warps of ground warp together to form a stiff surface at the back while the pile yarns stay at the front for a unique texture.
5. **Loop Formation**: As the Reed beats up the weft yarn, it pushes away the yarn of the pile warp onto the surface of the fabric. The patterns and design developments are made according to the required texture and various characteristics of the fabric properties.
6. **Emery rollers**: According to the end use, the softness and feathering are achieved by passing the fabric through the emery rollers (Fig. 11).

Terry Fabric has many characteristics and properties according to its end use, such as dish towels, baby products, towels, bathrobes, active wear, sweatshirts, wash clothes, bath mats, mattress protection, hoodies, bedding, etc. [32].

Fig. 11 Terry fabric (by Authors)

2.4.4 Fleece Fabric

Fleece is another fabric that involves high production in South Asian countries and demands advanced textile processing for manufacturing to construct innovative fabrics. The intricate tubular and double-layered fabrics are formed with the help of the mechanisms of circular knitting machines. The fleece fabric quality depends on the loop formation to develop characteristics like softness and stretch in the fabric. After the knitting process, the fabric then undergoes the process of brushing or raising to lift the fibres upwards to achieve a feathery, plush, and frizzy texture, which traps air in between the raised fibres to enhance the insulation properties.

The process of fleece fabric manufacturing is as follows:

1. **Circular Knitting Process**: The main pillar of making fleece is circular knitting. The yarn forms loops in the form of circular arrangements of needles and moves concurrently to continue the knitting process, forming a tube-shaped structure and hollow fabrics.
2. **Brushing/Raising Process**: After the formation of a double-layer hollow fabric, the next step is brushing the fabric, which helps raise the upper layer to intensify the fluffiness and hairiness of the fabric. During this process, the fabric is speedily passed through rotating brushes, which will lift the fibres from the surface.

Following are the properties of fleece fabric:

1. Soft texture
2. Warm Surface
3. Comfortable feel
4. Insulation property
5. Lightweight
6. Moisture-wicking nature (Fig. 12).

Fig. 12 Fleece fabric (by Authors)

It is very commonly used in apparel products like cold-resistant fabrics such as active wear, hoodies, and jackets. It is also being used in home textile products like bedding items, blankets, and throws. Other than this, it is used in outdoor applications in camping products, sportswear, and hiking gear. Beyond this, it is widely used in pet products, fulfilling both functional properties as well as comfort needs.

2.5 *Wearable Technology and Smart Textiles*

The world is adopting new technologies in the area of technical textile and wearable technology, involving smart solutions to problems. It marks a transformative era for traditional textiles to evolve with advanced functions and performances. Smart textiles are also known as e-textiles, with the world revolving around the integration of wearable electronics that heightened the ability of the fabric to react, sense, and communicate. Wearable textiles are a high-tech and complex technology for monitoring various aspects in the fields of medical, sports, and automation textiles. Other than smart textile innovation, technical textile is also accelerating in its domain with the introduction of diverse fabrics with enhanced functional properties and a variety of applications in different sectors such as health, infrastructure, automation, etc. Now, new experiments are being done to combine smart textiles and technical textiles together to increase the electronic sensory value as well as the improvement of functional properties. Globally, these fields of textiles are prospering daily, with new evidence of products catering to new challenges [33].

The research and development in these fields of textiles in south Asian countries like Pakistan, India, and Bangladesh is in progress, but at a very low altitude, with current experiments and ideas developed on a lab scale. Many innovative studies have been done in these states, but bulk-scale production is still a challenge that needs to be accomplished [34, 35] (Table 4).

2.6 *Industry 4.0*

Globally, the world has adopted the integration of Industry 4.0 with its involvement in textile technologies. The South Asian countries have also unfolded their pathways to revolutionise unique aspects of Industry 4.0 in their region and industry.

1. **Identification of Barriers and Collaboration**: India has formed its way towards Industry 4.0 by making several changes and understandings of its barriers. The research conducted in India shows the hurdles to the progress of industry, such as a lack of trained staff members, commitment issues, and limited resources from the government. To overcome these problems, a triple-helix model of collaborative framework has been introduced among the textile industry sectors,

Table 4 Pakistan, India, and Bangladesh innovation studies and developments (by Authors)

Country	Innovation	Description
Pakistan	Conductive textiles for wearable electronics	• Responsive stain sensor for wearable textiles and wireless connectivity [36] • Smart fabrics and temperature sensors [37] • Conductive smart textile cotton based non-woven composite of hydrogel alginate [38] • Wearable textile stitched technology based on piezoresistive sensors [39]
India		• Versatile and durable: innovations in reusable conductive textiles for wearable electro/physiological monitoring [40] • Powering the future: integrating RF rectenna and thin film solar cell for wearable electronics energy harvesting [41] • Multifunctional application of smart textiles 2-D MXene [42]
Bangladesh		• Performance of smart clothing for textile based antennas [43] • Sustainable IOT automation for Bangladesh garment industry [44]
Pakistan	Medical textiles	• Curcumin-infused polyurethane: a novel approach for antibacterial medical textile applications [45] • Exploring advanced fibres for medical textile applications [46]
India		• Nature's defence: chitosan-neem synthesis for antimicrobial medical textiles [47] • Revolutionizing healthcare: nonwoven innovations for advanced medical applications [48]
Bangladesh		• Medical textiles in Bangladesh: significance and future prospects [49] • Evaluating the effectiveness of antimicrobial textiles [50] • Ensuring safety in the development of medical textiles for female garment workers [51] • Knitting gauze fabric as advanced wound dressing for medical application [52]
Pakistan	Technical textiles	• Advancing defence with thermal camouflage polyester-cotton blend coated in graphene oxide [53] • Mosquito-repellent camouflage fabric infused with eucalyptus oil and moringa oleifera gum [54] • Advancing woven fabric density prediction through artificial neural networks in textile engineering [55]
India		• Strategic innovations: technical textiles redefining military applications [56]
Bangladesh		• Geotextiles, industrial applications in the realm of technical textiles [57]

including academia and government. This adoption aims to pave the way towards a smoother Industry 4.0 [58].

2. **Sustainable Textile Manufacturing**: Manufacturing South Asian countries are now moving towards sustainable development in the textile industry. Industry 4.0 emphasises an eco-friendly environment and ecological responsibility, along with sustainable practices in product design. Pakistan's denim industry is highly

advanced in opting for sustainable applications in its production process. Other than this, Bangladesh is also practicing this integration through the Sustainable Industrial Water Projection (SIWP) tool to manage the water waste and pollution caused by it [59].

3. **Retail Industry and its Impacts**: All three major countries—India, Pakistan, and Bangladesh—are on the verge of incorporating advanced technologies in the retail sector. The core industry 4.0 technologies, like cloud computing, 3D printing, analytical data, advanced software development, and robotics, are being utilised to enhance efficiency along with their competitive benefits [60].

All three countries have been prospering in the field of textiles according to the resources they have and moving forward with advanced strategies that assist them to confront their challenges. The navigation towards Industry 4.0 provides an understanding of a variety of opportunities for collaborative frameworks, infrastructure, textile production processes, strategic orientation, and sustainable integration, amplifying business intelligence systems. The South Asian region is not only the adoption of technology but also a unique transformation that involves the traditional core with a modern approach [61–66].

References

1. Lefferts HL (1992) Southeast Asian textiles: new research and writing. J Southeast Asian Stud 405–417
2. Janaway RC, Coningham RA (2010) A review of archaeological textile evidence from South Asia. South Asian Stud
3. Islam M, Khan AM, Islam M (2013) Textile industries in Bangladesh and challenges of growth. Res J Eng Sci
4. Allen RC (2009) The industrial revolution in miniature: the spinning Jenny in Britain, France, and India. J Econ Hist 69(4):901–927
5. Gekas S (2002) The organization of Indian textile technology before and after the European arrival
6. Roy T (2010) Technological change in Indian textiles industry, 1991–2006. Int J Technol Glob 5(1/2):124–131
7. Gvrk KA (2012) Emerging trends in textile industry: a supply chain perspective with special reference to India. Int J Mark, Financ Serv Manag Res 123–131
8. Nayak R, Padhye R (n.d.) Artificial intelligence and its application in the apparel industry. Autom Garment Manuf
9. Hahn B (2016) Spinning through the history of technology: a methodological note. Text Hist
10. Bellis M (2021) James Hargreaves and the invention of the Spinning Jenny
11. Bellis M (2022) Thoughtco. Accessed www.thoughtco.com: https://www.thoughtco.com/spinning-mule-samuel-crompton-1991498
12. Yin R, Tao X, Jasper W (2020) A theoretical model to investigate the performance of cellulose yarns constrained to lie on a moving solid cylinder. Cellulose
13. Ahmed S, Syduzzaman M, Ashique SM (2015) Comparative study on ring, rotor and air-jet spun yarn. Eur J Sci Res
14. Nemr AE (2012) From natural to synthetic fibers. Nova Science Publishers
15. Slootmaker JM (2010) Types of fibre. In: Müssig PD-I (ed) Industrial applications of natural fibres: structure, properties and technical applications, pp 41–48. https://doi.org/10.1002/9780470660324

16. Mehar MA (2021) Magnitude of investment and global value chain: a case study of textile and clothing industry of Pakistan. J Text Inst
17. Sanjay MR, Arpitha GR, Naik LL, Gopalakrishna K, Yogesha BJ (2016) Applications of natural fibers and its composites: an overview. Nat Resour 108–114
18. Dhakal PH, Ismail DO (2020) Potential of natural fibres and their composites for South Asian countries: moving towards sustainability. In: 6th SONEUK conference on engineering challenges and opportunities: post COVID-19
19. Md Syduzzaman MA (2020) Plant-based natural fibre reinforced composites: a review on fabrication, properties and applications. Coatings 1–34
20. Mazumdar D (1984) The issue of small versus large in the Indian textile industry an analytical and historical survey dipak mazumdar world bank staff working papers number 645 SWP645 public disclosure authorized public disclosure authorized public disclosure authorized publ. In: The international bank for reconstruction and development/the world bank
21. Bellis M (2020) Thoughtco. Accessed www.thoughtco.com: https://www.thoughtco.com/power-loom-edmund-cartwright-1991499
22. Bellis M (2020). Thoughtco. Accessed www.thoughtco.com: https://www.thoughtco.com/joseph-marie-jacquard-1991642
23. Schröter A, Gloy Y-S, Corves B, Schwarzfischer F (2016) Analysis of the weft insertion process and development of a relay nozzle concept for air-jet weaving. Res Gate
24. Gandhi K (2020) The fundamentals of weaving technology. In: Gandhi K (ed) Woven textiles (second edition): principles, technologies and applications. Woodhead Publishing, pp 167–270. https://doi.org/10.1016/C2017-0-01396-4
25. Gulveren H (2018) Circular knitting. Retrieved from knitting industry: https://www.knittingindustry.com/good-year-for-eu-circular-and-socks-knitting-machine-producers/
26. GmbH S (1993) Flat knitting machine. Retrieved from google art & culture: https://artsandculture.google.com/asset/flat-knitting-machine-stoll-gmbh/2QE_3dCXjjba2w
27. Banerjee PK (2006) Technological innovations in woven fabric manufacturing process. Indian J Fiber Text Res 125–133
28. Paul R (2015) Denim and jeans: an overview. In: Paul R (eds) Denim: manufacture, finishing and applications, pp 581–599. Woodhead Publishing
29. Raina MA, Gloy YS, Gries T (2015) Weaving technologies for manufacturing denim. In: Paul R (ed)Denim: manufacture, finishing and applications, pp 159–187. Woodhead Publishing
30. Denim production in Pakistan: Environmental Impacts, Strategies, and Future Directions (2023) Accessed Dress Ecode: https://dress-ecode.com/en/2023/04/13/denim-production-in-pakistan-environmental-impacts-strategies-and-future-directions/
31. Ahmad Y (n.d.) Textile industry of Pakistan. Report on textile industry of Pakistan
32. Ünal PG (2012) 3D woven fabrics. INTECH open access publisher
33. Schwarz A, Langenhove LV, Guermonprez P, Deguillemont D (2013) A roadmap on smart textiles. Taylor & Francis
34. Asiya Chaudhary MN (2012) Technical textiles in India: the trade perspective. JM Int J Manag Res 2(6):1–11
35. Zaman N, Memon NA (2007) Pakistan lags behind in technical textile. J Manag Soc Sci 3(2):120–127
36. Iqra Abdul Rashid IS (2022) Highly resilient and responsive fabric strain sensors: their effective integration into textiles and wireless communication for wearable applications. Sens Actuators: A Phys 1–11. https://doi.org/10.1016/j.sna.2022.113836
37. Abher Rasheed AI (2024) Design and integration of textile-based temperature sensors for smart textile applications. Smart Mater Struct 1–8. https://doi.org/10.1088/1361-665X/ad1c52
38. Shahood uz Zaman BM (2023) Development of conductive cotton non-woven alginate hydrogel composite for smart textiles. J Polym Environ 3998–4006. https://doi.org/10.1007/s10924-023-02847-5
39. Nauman Ali Choudhry AR (2020) Design, development and characterization of textile stitch-based piezoresistive sensors for wearable monitoring. IEEE Sens J 1–10

40. Kisannagar RR, Singh M, Gupta D (2021) Multifunctional, Wash durable and re-usable conductive textiles for wearable electro/physiological monitoring. Macromol Mater Eng
41. Naresh B, Singh VK, Sharma VK (2020) Integration of RF rectenna with thin film solar cell to power wearable electronics. Int J Microw Wirel Technol 1–12. https://doi.org/10.1017/S1759078720000410
42. Abbas Ahmed MM (2020) Recent advances in 2D MXene integrated smart-textile interfaces for multifunctional applications. Chem Mater 10296–10320
43. Khaleda Ali SA (2017) Design and performance analysis of a textile antenna suitable for smart clothing.In: IEEE region 10 humanitarian technology conference, pp 367–370
44. RA Ahmed KR (2023) An IoT based sustainable automation system for the garments industry in Bangladesh. In: Proceedings of international conference on planning, architecture and civil engineering, pp 962–967
45. Noureen Arshad KM (2022) Synthesis of novel curcumin-based aqueous polyurethane dispersions for medical textile diligences with potential of antibacterial activities. Polym Bull 79:7711–7727. https://doi.org/10.1007/s00289-021-03871-y
46. Ahmad S, Rasheed A (2020) Fibers for technical textiles. Springer Nature Switzerland AG
47. Revathi T, Thambidurai S (2017) Synthesis of chitosan incorporated neem seed extract (Azadirachta indica) for medical textiles. Int J Biol Macromol 1890–1896
48. Ajmeri IJR, Ajmeri CJ (2011) Nonwoven materials and technologies for medical applications. Woodhead Publishing Limited
49. Akter S, Azim AY, Al Faruque MA (2014) Medical textiles: significance and future prospect in Bangladesh. Eur Sci J 10(12):488-502
50. Mohammad Amir Hamzah AI (2015) A comprehensive analysis on the efficacy of antimicrobial textiles. Int J Text Sci 137-145. https://doi.org/10.5923/j.textile.20150406.02
51. Ahmed MT, Rasel M, Eshita I, Sadi MS (2018) An approach for development of medical textiles; safety issue of garments (female) worker
52. Alam SMM FM (2018) Development of knitted gauze fabric as wound dressing for medical application. Adv Res Text Eng 1–4
53. Suniya Tariq ZT (2022) J Nat Fibers 19(16):14222-14234. https://doi.org/10.1080/15440478.2022.2117759
54. Faiza Anwar MA (2023) Development of mosquito-repellent camouflage fabric using eucalyptus oil with moringa oleifera gum. ChemEngineering 1–12. https://doi.org/10.3390/chemengineering7040064
55. Haleem N, Rehman A (2011) Modeling the woven fabric density by using multiple linear regression and feed forward back propagation artificial neural networks. In: Proceedings of international Bhurban conference on applied sciences & technology
56. Revaiah RG, Kotresh TM, Kandasubramanian B (2019) Technical textiles for military applications. J Text Inst
57. Hasan R (2020) An overview of geotextiles: industrial application in technical textiles. J Text Sci Fash Technol
58. Ali S (2021) Operational readiness of Pakistan textile industry to implement industry 4.0 (a study of Hyderabad textile). Int J Eng Appl Sci Technol
59. Rathore DB (2023) Textile industry 4.0: a review of sustainability in manufacturing. Int J New Media Stud (IJNMS)
60. Sayem SM, Nahar N, Haque MS (2021) Industrial water management and sustainability: development of SIWP tool for textile industries of Bangladesh. Water Resour Ind
61. Baines (1835) www.handweaving.net Upload User:ClemRutter, Public Domain: https://commons.wikimedia.org/w/index.php?curid=6641264
62. Baines (n.d.) Spinning a revolution. Accessed https://commons.wikimedia.org/w/index.php?curid=6641264
63. Baswas A (2017) Rapier loom. Accessed Slideshare: https://www.slideshare.net/amitbiswas420/rapier-loom-75352015
64. Bellis M (2023) Thoughtco.com. Accessed www.thoughtco.com: https://www.thoughtco.com/industrial-revolution-in-pictures-1991940

65. Easy SS (n.d.) Types of fleece that you should know. Accessed so sew easy: https://so-sew-easy.com/types-of-fleece-that-you-should-know/
66. Mildason (n.d.) Colored denim fabric purchase price + quality test. Accessed Kohsrang: https://www.khoshrang-co.com/2023/03/04/colored-denim-fabric-purchase-price-quality-test/

Technological Gaps in Past and Present Times in South Asian Countries

Muhammad Hussnain Sethi, Zarwa Mughal, and Hafsa Jamshaid

Abstract The rich textiles of South Asian countries are intricately woven with the threads of history, tradition, and innovation. Among the numerous sectors that have played a pivotal role in shaping the cultural and economic landscape of the region, the textile industry stands out as a vibrant and dynamic force. This chapter unfolds a narrative that spans centuries, enlightening the technological milestones that have marked the journey of creative textile production in this diverse and culturally rich part of the world.

1 Introduction

The rich textiles of South Asian countries are intricately woven with the threads of history, tradition, and innovation. Among the numerous sectors that have played a pivotal role in shaping the cultural and economic landscape of the region, the textile industry stands out as a vibrant and dynamic force. This chapter unfolds a narrative that spans centuries, enlightening the technological milestones that have marked the journey of creative textile production in this diverse and culturally rich part of the world. The exploration begins with a retrospective gaze, delving into the past to unravel the foundations laid by our forebearers in textile and fabric manufacturing. From the artisanal craftsmanship of yesteryears to the industrial revolutions that transformed the landscape, we trace the historical trajectories that have shaped the

M. H. Sethi (✉)
School of Fine Arts, Design and Architecture, GIFT University, Gujranwala, Pakistan
e-mail: Hussnainsethi@yahoo.com; Hussnain.sethi@gift.edu.pk

School of Textile Science and Engineering, Zhejiang Sci-Tech University, Hangzhou, China

Z. Mughal
School of Design, Wuhan Textile University, Wuhan, China

H. Jamshaid
School of Engineering and Technology, National Textile University, Sheikhupura Road, Faisalabad, Pakistan

H. Jamshaid and A. Dad (eds.), *Creative Textile Industry*, SDGs and Textiles,
https://doi.org/10.1007/978-981-97-3802-1_6

creative textile industry across South Asian countries up to the turn of the twentieth century.

The chapter scrutinizes the present technological landscape, identifying gaps and disparities that have emerged in the wake of rapid advancements. Through an analytical lens, we navigate the challenges and opportunities presented by the integration of technology in fabric manufacturing, processing, and design methods. Yet, the scope extends beyond a mere retrospective and contemporary analysis. The chapter looks forward, engaging with the intriguing concept of Multi-Specie design processes and methods that hold promise for the future of creative textile industries in South Asian countries. By projecting our gaze towards the horizon, we aim to determine the potential challenges and prospects that lie ahead, providing a glimpse into the future of this dynamic and ever-evolving sector. This chapter is not only a scholarly endeavor for students, pedagogues, and academics but also a voyage for the curious general reader seeking to broaden their understanding of the rich tapestry of South Asian creative textile products. It is a step towards fulfilling the long-felt need for a comprehensive and up-to-date handbook, providing a foundation of knowledge for this integral sector and its allied industries. This chapter initiates by examining the technological distinctions between Britain and South Asia before delving into a comparative analysis of the textile industries in India, Pakistan, and Bangladesh. The historical technological differences, particularly rooted in the Industrial Revolution that unfolded predominantly in Britain, have had a profound impact on shaping diverse trajectories in textile manufacturing. The advancements during the Industrial Revolution, including mechanization and the utilization of steam engines, set the stage for significant disparities in manufacturing approaches between Britain and South Asian regions. Understanding these historical distinctions provides a crucial backdrop for comprehending the subsequent development of textile industries in the respective regions.

2 The Technological Landscape of Textile Industries: A Comparative Analysis of Britain and South Asia

2.1 Technological Integration in Britain's Textile Industry

Britain's textile industry is prominent for its technological innovation and efficacy. Advanced technologies, such as automated weaving and quality control systems, have significantly enhanced production rates, reduced labor costs, and minimized human errors. This adoption of cutting-edge technologies has not only strengthened Britain's position in the global textile market but also set a benchmark for technological excellence, making it a frontrunner in the industry.

2.2 *Diverse Technological Scenarios in South Asia*

The narrative that all of South Asia is behind in technology isn't accurate, especially when looking at the textile industries of India, Pakistan, and Bangladesh. Each of these countries has its own unique problems and strengths, which come from their different histories, cultures, and economic situations.

2.3 *India's Textile Industry: Innovation and Tradition*

India's textile sector is a vibrant mix of traditional craftsmanship and modern technological adoption. While certain areas rely on manual labor and age-old techniques, there's a growing emphasis on technological upgradation, especially in automated spinning and fabric processing. The government's support for technological innovation, as seen in the Technology Upgradation Fund Scheme (TUFS), has spurred improvements in efficiency and global competitiveness [1]. Furthermore, India's strength lies in its diverse product range, from exquisite handlooms to high-quality cotton and synthetic fabrics, catering to both domestic and international markets.

2.4 *Pakistan's Textile Dynamics: Challenges and Opportunities*

Pakistan's textile industry, being a major contributor to the country's economy, faces challenges in technological modernization due to various factors, including energy constraints [2]. However, the sector shows resilience and potential for growth, particularly in cotton production, where Pakistan stands as one of the leading producers globally [3]. Efforts towards incorporating sustainable practices and improving energy efficiency are gradually taking shape, with initiatives aimed at reducing water consumption and adopting cleaner production technologies [4].

2.5 *Bangladesh's Garment Industry: Rapid Growth Amidst Technological Evolution*

Bangladesh's garment industry, the backbone of its economy, has witnessed rapid growth and is actively working towards overcoming technological and environmental challenges. The focus on compliance with international labor and environmental standards has led to investments in greener technologies and better waste management practices. The adoption of automated cutting machines and efforts to improve energy efficiency reflect a commitment to technological advancement and sustainability [5].

2.6 *Addressing the Technological Divide*

The technological divide between Britain and South Asian countries highlights not only disparities but also opportunities for mutual learning and collaboration. While Britain can offer technological expertise and innovation, South Asian countries bring rich traditions, craftsmanship, and a vast market to the global textile landscape.

2.7 *Bridging the Gap Through Collaboration*

Encouraging technology transfer, joint ventures, and knowledge exchange programs between Britain and South Asian countries could accelerate technological adoption and innovation in the latter's textile industries. Focusing on sustainable and environmentally friendly ways of making textiles can also improve their standing in the world market and their commitment to taking care of the environment.

2.8 *Leveraging Unique Strengths*

Each region's unique strengths Britain's technological advancements and South Asia's rich textile heritage can be leveraged to foster a more inclusive and sustainable global textile industry. For instance, integrating traditional designs with modern fabrics and production techniques can create high-value products appealing to diverse markets.

2.9 *Towards a Sustainable and Technologically Advanced Future*

The future of the textile industry lies in balancing technological advancement with environmental sustainability and social responsibility. Encouraging the adoption of green technologies, efficient resource use, and fair labor practices are critical steps towards achieving this balance. By recognizing and addressing the unique challenges and strengths of Britain and South Asian countries, the global textile industry can move towards a more equitable, innovative, and sustainable future.

3 Historical Perspective; Fabric Manufacturing in South Asia

3.1 Weaving Tradition and Modernity

Continuing our exploration through the historical loom of South Asian textile heritage, we now discuss the evolution of fabric manufacturing techniques within the intricate landscapes of India, Pakistan, and Bangladesh. Each stitch in this narrative tells a story of modernization, adaptation, and the outstanding interfusion between tradition and progress.

3.1.1 India: Pioneering Craftsmanship to Industrial Revolution

In the Indian subcontinent, the evolution of fabric manufacturing techniques is a testament to centuries of pioneering craftsmanship. From the royal courts adorned with hand-spun cotton and silk to the flourishing trade routes that disseminated Indian textiles globally, novelty has been entangled into the very fabric of the nation's heritage. Technological advancements during the Mughal era, elevated the scale of production, solidifying India's status as a global hub for finely crafted fabrics.

The impact of historical events, notably British colonization and subsequent industrialization, ushered in a transformative phase. While traditional methods endured, the introduction of power looms and mechanized spinning marked the dawn of modernization in the Indian textile sector. The marriage of tradition and innovation became the hallmark of India's fabric manufacturing evolution. The fabric manufacturing techniques of India have, over the centuries, undergone a remarkable metamorphosis. In ancient times, skilled artisans engaged in laborious hand-spinning and hand-weaving processes, producing textiles that were not only aesthetically pleasing but also highly prized. This craftsmanship was not confined to local use but found its way into the global trade networks through well-established routes. The Mughal era marked a significant period of progress. The intricate designs and vibrant colors of Indian textiles became sought after across continents, creating a global demand that fueled the growth of the industry.

However, the colonial period brought about a seismic shift in the fabric manufacturing landscape. British rule, established in the nineteenth century, brought industrialization to India. Traditional handcrafts faced competition from mass-produced textiles, as power looms and mechanized spinning machines were introduced. While this period marked a departure from artisanal methods, it also paved the way for increased production and economic growth.

The chemistry between traditional techniques and modern approaches has been a defining characteristic of India's fabric manufacturing journey. Even today, amidst the whir of advanced machinery, handloom weaving perseveres, distinguished for its craftsmanship and cultural significance. The merging of traditional and modern methods in textile production demonstrates a subtle approach, amalgamating heritage with efficiency.

3.1.2 Pakistan: A Quilt Woven with Tradition and Technological Marvels

In Pakistan, the historical journey of fabric manufacturing discloses as a dynamic relationship between tradition and technological marvels. The traditional art of handloom weaving, firmly rooted in cultural identity, seamlessly blends with technological advancements [6]. The acceptance of contemporary spinning mills and the integration of synthetic fibers catapulted Pakistan into a global textile arena, illustrating the country's ability to balance tradition with progress.

The fabric manufacturing techniques in Pakistan have a rich history. The craft of handloom weaving, passed down through generations, reflects the cultural identity of the region. This traditional craftsmanship, characterized by intricate designs and manual techniques, has not only preserved cultural heritage but also contributed to the distinctiveness of Pakistani textiles. However, as the global textile landscape evolved, Pakistan embraced technological advancements to stay competitive. The adoption of advance yarn manufacturing mills and the integration of synthetic fibers marked a significant shift [7]. This transition allowed for increased efficiency, higher production volumes, and a broader range of textile products. The impact of historical events, such as the partition in 1947, posed challenges but also served as a crucible for resilience and adaptation within the fabric manufacturing sector. The outcome is a strong industry, standing strong despite changes and moving forward by combining traditional skills with modern technology. After the partition, the fabric-making part of Pakistan changed a lot. Skilled workers moved from India to the new country, bringing their knowledge. This mix of talent, along with a promise to modernize, set the foundation for a successful textile industry.

The fusion of advanced machinery, including automated looms and computerized knitting machines, enabled Pakistan to diversify its textile offerings. The production of synthetic fibers added a new dimension to the industry, expanding the range of materials and products. While traditional techniques persisted, the embrace of modern technology positioned Pakistan as a major player in the global textile market (Fig. 1).

3.1.3 Bangladesh: Jute Mills to Global Leadership

In the agrarian landscape of Bangladesh, the transformation from traditional jute cultivation to industrial scale production marked the beginning of its journey in fabric manufacturing. The establishment of jute mills in the nineteenth century laid the groundwork for what would become a significant sector of the Bangladeshi economy. Despite facing immense challenges during the Bangladesh Liberation War in 1971, the resilience of the textile sector was unmistakable. Post-independence, Bangladesh

Fig. 1 Work of an artisan on a handloom in Pakistan

embarked on a journey to diversify and modernize its textile and garment industry, leveraging its low labor costs and favorable trade agreements to attract foreign investment [8]. By the 1980s and 1990s, Bangladesh had begun to make its mark on the global stage, with the Ready-Made Garment sector experiencing exponential growth [9]. This growth was partly due to the Multi-Fibre Arrangement of the 1970s, which allowed developing countries like Bangladesh to have quota-free access to the US market, and the subsequent phase-out of the Multi-Fibre Arrangement in 2005, which did not diminish Bangladesh's competitive edge as feared but rather saw a continuation and acceleration of growth [10, 11]. Today, Bangladesh is the second-largest apparel exporter in the world, trailing only behind China [12, 13]. Its garment industry accounts for over 80% of the country's total export earnings and employs millions, making it a critical driver of Bangladesh's economic development [14]. The country's ascendancy to global leadership in textile and garment manufacturing is evidenced by its significant market share in the European Union and the United States, along with a growing presence in other markets. Moreover, Bangladesh's commitment to sustainability and ethical manufacturing practices, as seen through initiatives like the Accord on Fire and Building Safety in Bangladesh, further strengthens its position as a global leader in the textile and garment manufacturing industry [15]. These developments underscore Bangladesh's remarkable journey from a jute-dominated economy to a powerhouse in global textile and garment production, showcasing the nation's adaptability, resilience, and strategic leveraging of its competitive advantages in the post-Liberation era. The incorporation of technology, from advanced spinning methods to eco-friendly processing techniques, reflects a commitment to sustainable and innovative fabric manufacturing practices. The evolution of fabric manufacturing in Bangladesh echoes a narrative of transformation, resilience, and a forward-looking approach. The fabric manufacturing techniques in Bangladesh have been shaped by a deep connection to the land and a history steeped in agricultural practices. Jute, known as the "Golden Fiber," has been a significant contributor to the country's textile heritage. Traditionally used for rope and twine, jute found its place in the global market with the establishment of jute mills in the nineteenth century. The jute industry became a major economic driver for Bangladesh, contributing significantly to its GDP [16]. However, the sector faced challenges during and after the Bangladesh Liberation War in 1971. The conflict disrupted operations, and the subsequent independence of Bangladesh led to the nationalization of jute mills. Despite these challenges, the textile sector displayed remarkable resilience and adaptability. In the latter half of the twentieth century, Bangladesh witnessed a diversification of its textile industry [17]. While jute remained an important component, the focus expanded to include cotton and synthetic fibers. Technological advancements, such as the adoption of modern spinning techniques and the implementation of eco-friendly processing practices, signaled a commitment to evolution and sustainability.

The integration of modern technology brought efficiency and precision to fabric manufacturing in Bangladesh. Bangladesh's textile and garment industry flourished, earning the country a reputation as a global leader in apparel exports. The fabric manufacturing journey of Bangladesh reflects a trajectory from traditional practices rooted in agriculture to a diversified, technology-driven industry. The resilience displayed during challenging periods and the forward-looking approach to sustainable practices underscore Bangladesh's significance in the global textile landscape.

This exploration of fabric making in India, Pakistan, and Bangladesh shows the detailed patterns stitched into the history and culture of these countries. From old-fashioned delicate craftsmanship to using modern technology, time keeps telling a story of new ideas and adjustments in the changing South Asian textile industry.

4 Advances in Dyeing and Printing

4.1 The Chromatic Canvas of Tradition and Innovation

Our exploration into the South Asian textiles world now bring us to the vibrant array of dyeing and printing methods in India, Pakistan, and Bangladesh. The historical canvas of these nations is adorned with the shades of ancient customs, while advancements in technology have steered in a dynamic and diverse range of opportunities in the field of dyeing and printing.

4.1.1 India: Chromatic Legacies of Cultural Brilliance

In India, the historical practices of dyeing and printing are deeply connected with the country's rich cultural heritage. The ancient art of tie-dyeing, known as Bandhani, has adorned fabrics for centuries, creating elegant patterns through the meticulous tying and dyeing of fabrics [18]. Additionally, the renowned art of block printing has been a backbone, employing exquisitely wooden blocks transferring vibrant pigments onto fabrics (Fig. 2).

Technological advancements in dyeing and printing processes in India have been a gradual evolution. Traditional methods have endured, cherished for their cultural significance, but modernization has brought efficiency and scale. The adoption of rotary screen printing, digital printing technologies, and eco-friendly dyeing processes has expanded the horizons of India's textile industry.

In the contemporary landscape, India stands at the intersection of tradition and technology in dyeing and printing. The harmonious blend of handcrafted methods and automated processes demonstrates a dedication to upholding cultural traditions while also embracing the efficiencies provided by state-of-the-art technology (Fig. 3).

4.1.2 Pakistan: A Canvas Crafted with Pigments and Precision

Pakistan's historical practices of dyeing and printing bear witness to a vibrant canvas crafted with pigments and precision. The art of Ajrak block printing, characterized by its distinct geometric patterns and earthy tones, has deep roots in the Sindhi culture [19]. Additionally, tie-dyeing techniques which is also used in India, such as Bandhani and Shibori, have flourished, contributing to the diversity of Pakistan's textile traditions. The technological strides in dyeing and printing processes reflect a commitment to precision and scalability. The introduction of modern printing machinery, including flatbed and rotary screen printers, has allowed for intricate designs and vibrant color palettes. Moreover, the utilization of reactive dyes and digital printing technologies has enhanced the precision and variety of patterns in the textile industry.

Fig. 2 Artisans are creating tie-dyeing patterns

In the ever-evolving technological landscape, Pakistan's dyeing and printing sector is undergoing continual transformation. Embracing the paradigm of automation has become integral for achieving large-scale production, a strategic move that not only enhances efficiency but also upholds uncompromised quality standards. This seamless integration of cutting-edge automation with traditional craftsmanship not only signifies adaptability but also positions Pakistan as a key player on the dynamic stage of the global market (Fig. 4).

This synergy of time-honored artistic expressions and state-of-the-art methodologies propels Pakistan into a realm of advancement and competitiveness. The sector's commitment to modernization not only ensures the streamlining of processes but also serves as a testament to the nation's ability to balance heritage with contemporary demands. As Pakistan continues to tread the path of technological advancement, its dyeing and printing industry emerges not merely as a participant but as a trailblazer, showcasing a harmonious marriage of tradition and cutting-edge techniques on the international stage.

Fig. 3 Screen prepared for screen printing

4.1.3 Bangladesh: A Canvas of Sustainability

Bangladesh, with its commitment to sustainable practices, has seen significant strides in the historical practices of dyeing and printing. Traditional techniques, such as Batik and Nakshi Kantha, reflect the cultural identity of the nation. These handcrafted methods have been complemented by modern techniques that prioritize eco-friendly processes and materials. Technological advancements in dyeing and printing in Bangladesh have been marked by a focus on sustainability. The adoption of water-saving dyeing techniques, eco-friendly pigments, and digital printing technologies has positioned Bangladesh as a leader in environmentally conscious textile production [20]. Additionally, the utilization of organic dyes and the implementation of energy-efficient machinery underscore the industry's commitment to a greener future. In the current state of technology, Bangladesh continues to champion sustainability in dyeing and printing practices. The industry has become a pioneer in adopting water recycling systems and investing in research for further eco-friendly alternatives. This holistic approach aligns with global trends towards environmentally responsible manufacturing.

Fig. 4 Different tie-dyeing patterns

4.2 *Comparative Analysis: Tradition, Technology, and Global Impact*

A thorough comparison of the historical practices, technological improvements, and current state of dyeing and printing in India, Pakistan, and Bangladesh reveals a rich landscape. Each country has embraced its cultural heritage, nurturing traditional techniques that have endured over time. Simultaneously, the infusion of modern technologies has propelled these nations to the forefront of the global textile industry.

India, known for its detailed tie-dye and block printing traditions, skillfully balances the old and the new, creating a vibrant fabric that resonates globally. Pakistan's precision in Ajrak block printing and adoption of modern machinery showcases a harmonious blend of heritage and efficiency. Bangladesh, a leader in sustainable practices, weaves a narrative of modernization aligned with contemporary environmental concerns.

As we examine the current technological landscape in dyeing and printing across South Asian countries, a common thread emerges the commitment to preserving cultural identities while adapting to the demands of a globalized market. This exploration through the chromatic canvas of tradition and innovation offers a glimpse into the artistry defining the South Asian textile industry. From historical legacies to cutting-edge technologies, the dyeing and printing practices in these nations contribute to the ever-evolving narrative of global textile excellence.

5 Garment Manufacturing: Past and Present

The journey through South Asian textile now takes us to the heart of the garment manufacturing realm. In the historical loom of India, Pakistan, and Bangladesh, the evolution of garment production has been marked by craftsmanship, adaptation, mixture between tradition and technology (Fig. 5).

Fig. 5 Workers are using Juki Sewing Machines to stitch garments in a South Asian industry

5.1 *Historical Development: Crafting Apparel Through Time*

Garment manufacturing in the South Asian context has a rich history of each nation. From the finely tailored garments of royal courts to the exquisite hand-embroidered fabrics, the historical development of garment manufacturing showcases the region's commitment to sartorial excellence [21]. In India, the ancient art of handloom weaving and intricate embroidery techniques laid the foundation for a tradition of garment craftsmanship that extends back to centuries. The Mughal era witnessed the emergence of skilled tailors and intricate garment designs, setting the stage for a rich legacy that persists to this day.

Similarly, Pakistan's garment manufacturing history reflects a fusion of traditional craftsmanship and cultural diversity [22]. The intricate detailing of traditional Pakistani attire, such as the shalwar kameez and sarees, exemplifies a meticulous approach to garment creation. Over time, this craftsmanship evolved to meet the demands of changing fashion trends and global markets. Bangladesh, known for its garment industry's meteoric rise, has a historical backdrop rooted in the production of traditional attire like the saree and kurta. The transition to mass garment production began in the mid-twentieth century, laying the groundwork for Bangladesh's emergence as a global hub for ready-made garments (Fig. 6).

5.2 *Technological Gaps: Stitching Together Past and Present Challenges*

The technological evolution in South Asian garment manufacturing is shaped by a complex blend of historical factors. A closer look at the origins of technological gaps in India, Pakistan, and Bangladesh reveals a story deeply connected to colonial legacies, labor dynamics, industrial fragmentation, and limitations in research and development.

5.2.1 Colonial Legacy: A Persistent Knowledge Gap

The colonial legacy casts a long shadow on the technological trajectory of South Asian garment manufacturing. The historical reliance on British mills during the colonial era left an enduring knowledge gap. The domination of external expertise hampered the development of indigenous technologies, hindering the emergence of a self-sufficient industry. The imprint of this colonial legacy persists, influencing how these nations engage with and adopt modern technological advancements.

Fig. 6 A worker is performing machine embroidery in Gujranwala, Pakistan

5.2.2 Focus on Cheap Labor: Post-independence Priorities

Post-independence, the pursuit of low-cost production became a central tenet in the economic strategies of India, Pakistan, and Bangladesh. This focus, while initially facilitating competitiveness in the global market, inadvertently discouraged significant investments in high-tech machinery and the development of advanced skills [23]. The emphasis on cost-effectiveness over technological breakthrough resulted in a prolonged technological lag, impacting the overall modernization of the garment manufacturing sector.

5.2.3 Fragmented Industry: Impediment to Large-Scale Upgrades

The South Asian garment manufacturing industry is comprised of numerous small and medium-sized enterprises. While this diversity contributes to the richness of the industry, it also poses a challenge for large-scale technological upgrades. The fragmented nature of the sector makes it difficult and expensive for individual enterprises to invest in cutting-edge machinery or adopt comprehensive technological solutions. As a result, the industry is characterized by a patchwork of technological advancements rather than a cohesive, synchronized evolution.

5.2.4 Limited R&D: The Struggle

The dearth of investment in research and development (R&D) stands as a formidable barrier to technological progress in the South Asian garment manufacturing sector. The limited allocation of resources to R&D initiatives stifles innovation and perpetuates a reliance on imported technologies. The absence of a robust R&D ecosystem hampers the ability of these nations to develop tailor-made solutions, leaving them susceptible to external market forces and hindering their capacity to carve out distinctive technological identities.

Addressing these multifaceted challenges requires a holistic approach that encompasses policy reforms, targeted investments, and collaborative efforts within the industry. The journey to bridge the technological gaps in India, Pakistan, and Bangladesh involves not only acknowledging historical influences but also charting a course towards a more technologically empowered and self-sufficient future for the garment manufacturing sector in South Asia.

5.3 Current State of Garment Manufacturing Technology

Despite the persistent challenges outlined in the previous section, the landscape of garment manufacturing technology in South Asia is far from static. In recent years, technology has surfaced in the garment manufacturing sector, accompanied by the integration of automation, digitization, and sustainability practices. This shift reflects a growing emphasis on technological advancements, streamlined processes, and eco-friendly initiatives within the industry.

5.3.1 India

Embracing a diverse approach, India mixes advanced technologies like automated cutting machines, computerized pattern making, and even 3D printing alongside handcrafted techniques like intricate embroidery [24]. This unique blend allows it to cater to a wide range of markets, from budget-conscious consumers to luxury fashion houses. The synchronicity of handloom weaving and modern production lines remains a defining feature of the Indian garment industry, contributing to its unique charm and appeal.

5.3.2 Pakistan

Pakistan has undergone significant advancements, adopting computerized knitting machines, digital printing, and automated quality control systems. This, coupled with adherence to quality standards, has positioned it as a competitive player in the global apparel market. The industry's capacity for large-scale production and focus on precision has attracted major international brands, further propelling its technological evolution.

5.3.3 Bangladesh

Renowned as a major player in the global ready-made garment industry, Bangladesh has harnessed technology not only for efficiency but also for sustainability. The implementation of digital supply chain management, automated cutting systems, and eco-friendly dyeing processes demonstrates its commitment to both modern technology and environmental responsibility. These efforts ensure continued relevance in the global market and position Bangladesh as a leader in sustainable garment manufacturing practices.

5.3.4 Beyond National Boundaries: Fostering Collaboration and Knowledge Sharing

While individual nations are making their own way of technological advancements, there is also a growing recognition of the benefits of regional collaboration and knowledge sharing. Initiatives like the South Asian Association for Regional Cooperation (SAARC) Textile Ministers' Meeting aim to foster cross-border cooperation in areas like technology transfer, R&D, and skill development [25]. Such efforts hold immense potential to accelerate the collective growth and technological progress of the South Asian garment manufacturing sector (Fig. 7).

5.3.5 Bridging the Gap: Future Way

It is clear that the technological landscape of South Asian garment manufacturing is undergoing a profound transformation. While historical factors and past challenges continue to cast their shadows, the region is witnessing a dynamic relationship of novelty, adaptation, and collaboration. Looking toward the future, all three countries are considering several key factors to shape the next phase of technological development:

Continued integration of advanced technologies: Automation, artificial intelligence, and robotics are poised to play an increasingly significant role in streamlining production processes and enhancing efficiency [26].

Fig. 7 Commonly used overlock sewing machine in South Asia

Focus on sustainability: Embracing eco-friendly practices, renewable energy sources, and resource-efficient technologies will be crucial for ensuring long-term sustainability and ethical production [27].

Investment in R&D: Fostering a vibrant R&D ecosystem, both at the national and regional levels, will be essential for developing indigenous technologies and fostering a culture of innovation.

Skill development and workforce adaptation: Equipping workers with the necessary skills to operate and manage advanced technologies will be key to bridging the skills gap and ensuring a smooth transition into the future of garment manufacturing.

6 Challenges in the Present Context

6.1 Navigating Technological Gaps and Contemporary Hurdles

In delving into the contemporary landscape of the creative textile industry in South Asian countries, it becomes evident that despite remarkable strides, existing technological gaps pose challenges to seamless innovation integration. Identifying and addressing these gaps is crucial for the sustained growth of the industry in the global market.

6.2 Existing Technological Gaps: A Critical Examination

The creative textile sector's technological gaps manifest in various facets of the production process. These gaps may arise from limited access to cutting-edge machinery, inadequate infrastructure, or a lag in the adoption of digital technologies. The coexistence of traditional handcrafted methods with modern industrial processes creates disparities in efficiency, posing challenges for streamlined production.

India, Pakistan, and Bangladesh each grapple with distinct technological gaps. For instance, India may face challenges in providing widespread access to digital design tools and automated machinery for small-scale artisans. The traditional yet intricate handloom weaving industry coexists with modernized methods, creating a diverse yet complex landscape. Pakistan, while technologically advanced, may encounter gaps in skills training and the integration of Industry 4.0 technologies [28]. Bangladesh, a global leader in ready-made garments, may need to address technological gaps in sustainability practices and digital supply chain management.

6.3 Adoption and Implementation Challenges: A Regional Perspective

The challenges faced by South Asian countries in adopting and implementing advanced technologies in the creative textile industry are multifaceted. Economic factors, such as the cost of technology acquisition and maintenance, can be prohibitive for smaller enterprises. Social factors, including resistance to change and the need for skills development, may impede the smooth transition to modernized production methods.

India, with its diverse textile landscape, may encounter challenges in standardizing technology across a spectrum of traditional and modern practices. The economic disparity between larger manufacturers and smaller artisans can create a digital divide. Pakistan may need to focus on upskilling its workforce to align with rapidly evolving technological trends, ensuring that the benefits of advanced technologies are inclusive. Bangladesh, while excelling in garment manufacturing, may face challenges in extending technology adoption to the broader textile creative sector, especially to smaller workshops.

6.4 Influencing Factors: Economic, Social, and Policy Dynamics

Several economic, social, and policy factors influence the adoption of advanced technologies in the creative textile industry. Economic factors include the availability of funding for technology investments and the overall economic stability of the region. Social factors involve the readiness of the workforce to embrace technological changes and the cultural significance attached to traditional craftsmanship. Policy dynamics, including regulatory frameworks and government incentives, play a crucial role in shaping the technological landscape of the industry.

The economic prosperity of South Asian countries influences their ability to invest in and adopt advanced technologies. Social dynamics, such as the preservation of cultural heritage and artisanal skills, may sometimes clash with the imperative for technological modernization. Policy frameworks that encourage novelty, provide financial incentives, and address infrastructural needs can significantly impact the industry's technological trajectory.

7 Future Prospects and Challenges

As we cast our gaze towards the horizon of the creative textile industry in South Asian countries, envisioning the future requires understanding of potential technological advancements and the challenges that lie ahead. The concept of Multi-Specie design processes emerges as a focal point, promising innovative possibilities but also posing its unique set of challenges.

7.1 Potential Advancements in Technology: Weaving Tomorrow's Fabric

The future of creative textile manufacturing holds exciting possibilities in technological advancements. Artificial Intelligence (AI), augmented reality, and 3D printing are poised to revolutionize design processes, enabling unparalleled levels of customization and creativity [29]. Sustainable technologies, including eco-friendly dyeing methods and recycled materials, are likely to take center stage as the industry places an increased emphasis on environmental responsibility.

India, with its diverse creative textile landscape, may witness a convergence of traditional craftsmanship with cutting-edge technologies. The integration of AI-driven design tools and sustainable production methods could redefine the industry's trajectory. Pakistan, as a technologically advanced player, may lead in the adoption of Industry 4.0 technologies, ensuring precision and efficiency. Bangladesh, with its focus on sustainability, may pioneer innovative eco-friendly practices and further establish itself as a global leader in responsible textile manufacturing.

7.2 *Multi-specie Design Processes: Shaping the Future Landscape*

7.2.1 Technological Gaps for a Flourishing Future

The concept of Multi-Specie design processes introduces a paradigm shift in creative textile manufacturing, particularly within the rich biodiversity context of South Asian countries. This innovative approach, leveraging the symbiotic relationships between different biological organisms, such as bacteria and fungi, in the textile design and production process, heralds a new era of sustainable manufacturing practices. It promises not only to enhance the ecological footprint of textile production but also to explore new textures, colors, and functional properties inherent in biological materials.

7.2.2 Relevance to South Asian Textile Manufacturing

South Asia's textile industry, rich in tradition and biodiversity, presents a fertile ground for integrating Multi-Specie design processes. The region's historical reliance on natural dyes and materials aligns with the principles of biofabrication, making it a promising candidate for pioneering sustainable and innovative manufacturing techniques. For instance, the use of microorganisms for dyeing fabrics not only reduces environmental harm but also revives ancient practices with a modern twist, offering textiles with unique properties such as improved durability and colorfastness.

7.2.3 Technological Gaps for a Flourishing Future

Recent projects and research within South Asian countries have demonstrated the practicality of integrating living organisms into textile production. For example, a collaborative initiative between Indian textile researchers and biotechnology firms has successfully developed bio-fabrics using yeast and bacteria, showcasing the potential for scalable production. Such endeavors underline the technical feasibility and economic viability of Multi-Specie design in the region's textile industry.

7.2.4 Potential Impact

The adoption of Multi-Specie design processes promises to revolutionize the South Asian textile sector by offering sustainable alternatives to chemical-intensive practices. Beyond environmental benefits, this approach can enhance the aesthetic and functional qualities of textiles, opening new markets and opportunities for innovation. Moreover, it aligns with global sustainability goals, positioning South Asia as a potential leader in eco-friendly textile manufacturing.

7.2.5 Challenges and Solutions

Despite its promising prospects, the implementation of Multi-Specie design faces several challenges, including the need for specialized knowledge, the development of supporting infrastructure, and market acceptance. Overcoming these obstacles requires concerted efforts from governments, industry stakeholders, and academia to foster an enabling ecosystem. Key strategies might include investing in research and development, providing training programs for artisans and manufacturers, and launching awareness campaigns to educate consumers about the benefits of biofabricated textiles.

Conclusion

As South Asian countries navigate the future of textile manufacturing, embracing Multi-Specie design processes offers a pathway to sustainable innovation. By building on the region's rich heritage of textile craftsmanship and its biodiversity, there is a unique opportunity to redefine the industry's environmental footprint while unlocking new creative potentials. With appropriate support and collaboration, Multi-Specie design can transition from a promising concept to a mainstream manufacturing practice, contributing to the global competitiveness of South Asia's textile industry.

8 Conclusion

8.1 Technological Gaps for a Flourishing Future

In traversing the complex tapestry of South Asian creative textile industries, our journey through the historical evolution, present challenges, and future prospects has unveiled a narrative of advancement, resilience, and cultural richness. As we conclude this exploration, it is imperative to distill the key findings surrounding technological gaps in fiber manufacturing, fabric production, dyeing and printing processes, and garment manufacturing. The synthesis of these insights not only illuminates the challenges faced by India, Pakistan, and Bangladesh but also underscores the critical importance of addressing these gaps for the sustained growth and flourishing future of the creative textile industry in the region.

8.2 Technological Gaps in Fiber Manufacturing

Our reviewing look into the historical perspective of fiber manufacturing revealed the rich legacies of craftsmanship in India, Pakistan, and Bangladesh. Traditional practices, deeply embedded in the cultural fabric, coexist with modern technological advancements. However, the journey is not without its challenges.

India, with its traditional hand-spun cotton and silk threads, faced challenges during the industrial revolution, introducing technological gaps between artisanal and mechanized production. Pakistan's handloom weaving, deeply rooted in cultural identity, adapted to modernization, introducing technological advancements but also creating disparities in skill levels. Bangladesh's transformation from traditional jute cultivation to modern technology adoption exemplifies resilience but not without the challenge of adapting to rapid changes.

8.3 Technological Gaps in Fabric Manufacturing

The evolution of fabric manufacturing techniques, as explored in our narrative, showcases a dynamic interplay between tradition and modernity. The technological gaps in fabric manufacturing stem from the coexistence of traditional handcrafted methods and advanced machinery.

India's diverse fabric manufacturing landscape grapples with integrating traditional craftsmanship with modern techniques. The challenge lies in ensuring that both approaches coexist harmoniously, offering a spectrum of choices to consumers. Pakistan, a dynamic player in the global textile market, faces challenges in upskilling its workforce to bridge the technological gap between traditional handloom weaving and modern spinning mills. Bangladesh's commitment to sustainable and innovative fiber manufacturing practices comes with the challenge of balancing tradition with modern eco-friendly processes.

8.4 Technological Gaps in Dyeing and Printing

Advances in dyeing and printing, a vibrant segment of the creative textile industry, present both opportunities and challenges. Traditional practices, deeply rooted in culture, must integrate with modern technologies for sustainable growth.

India's historical tie-dyeing and block printing traditions coalesce with modern advancements, marking a significant evolution in the region's textile craftsmanship. This integration, evident in the use of digital design and laser-cutting in block printing, allows for intricate designs and faster production. Similarly, in Pakistan, the Ajrak printing process benefits from automated dyeing techniques, aiming for consistency and efficiency in its famed intricate patterns.

However, this fusion of tradition with technology introduces challenges, notably the potential loss of traditional skills as artisans lean towards modern methods. Standardizing these advancements across India and Pakistan's diverse practices is daunting, requiring significant investment and innovation tailored to each unique tradition. The environmental impact of modern techniques and the economic barriers to accessing new technologies further complicate this transition. Despite these obstacles, the modernization of traditional crafts opens up opportunities. There is a burgeoning global demand for artisanal and sustainable products, and digital platforms provide artisans with unprecedented access to international markets. To navigate these technological gaps, collaborative efforts from governments, NGOs, and the private sector are crucial. They can offer support through technology access, training, and financial assistance, fostering an environment where innovation complements tradition. Balancing these aspects is essential for the sustainable development of the textile sector in India, Pakistan, and Bangladesh, ensuring that these age-old crafts not only survive but thrive in the modern era.

8.5 Technological Gaps in Garment Manufacturing

In the South Asian garment manufacturing landscape, transitioning from traditional craftsmanship to modernized production showcases both the promise of innovation and the profound challenges of equitable technology adoption. India, with its storied history in textile artisanship, faces significant hurdles in making technology accessible across its garment industry. The disparity between large-scale manufacturers and small to medium-sized enterprises (SMEs) is stark, with SMEs often struggling to secure the necessary capital for embracing advanced technological solutions such as automated cutting machines, computer-aided design (CAD) systems, and 3D modeling for garment design. This gap in technology adoption not only limits productivity and market access for these entities but also highlights broader inclusivity issues within India's technological evolution in the garment sector.

Similarly, Pakistan's garment sector, despite integrating modern technology, faces the challenge of harmonizing skills training with technology integration. This situation is exacerbated by economic struggles and electricity supply issues, hindering the consistent use of advanced technologies. Additionally, the level of education among the labor force in India, Pakistan, and Bangladesh often does not equip workers with the foundation necessary to adopt and adapt to new technological tools and processes effectively. The skills and education gap presents a significant barrier to leveraging technological advancements in garment manufacturing fully.

In Bangladesh, the garment industry, crucial to the country's economy, also navigates the dual challenges of modernizing production while ensuring the inclusivity of technology adoption. Smaller workshops, integral to the nation's garment production ecosystem, often find it difficult to access and implement new technologies due to similar constraints of capital and skilled labor. This underscores the need for targeted interventions that support technology dissemination and workforce training at all levels of the industry.

To address these multifaceted challenges, a comprehensive strategy involving policy reform, targeted investments in technology and skills training, and collaborative efforts between governments, industry stakeholders, and educational institutions is essential. Such efforts must aim to bridge the technological and educational gaps, fostering an ecosystem where traditional craftsmanship and modern technology coalesce. By prioritizing technological upskilling and inclusivity in advancements,

Fig. 8 Ajrak -Created through the art of Block Printing

South Asia can pave the way toward a more resilient, competitive, and sustainable garment manufacturing sector that not only preserves its rich heritage but also embraces the future's potential (Fig. 8).

8.6 Emphasizing the Importance of Addressing Technological Gaps: A Call to Action

The comprehensive exploration of technological gaps across fiber, fabric, dyeing and printing, and garment manufacturing underscores the urgency of addressing these challenges for the future of the creative textile industry in South Asian countries. The integration of technology is not merely a matter of industry progression; it is a crucial step towards preserving cultural heritage, sustaining economic growth, and fostering environmental responsibility.

Addressing technological gaps ensures the preservation of traditional craftsmanship, preventing the erosion of invaluable cultural practices. It empowers artisans with new tools and techniques, enabling them to thrive in a modernized industry landscape. The integration of advanced technologies ensures the competitiveness of

South Asian creative textile products in the global market, fostering economic growth and creating employment opportunities.

Moreover, technology plays a pivotal role in promoting sustainability. From eco-friendly dyeing methods to the implementation of digital supply chain management, addressing technological gaps allows the industry to embrace environmentally conscious practices. This not only aligns with global trends but also positions South Asian countries as leaders in responsible textile manufacturing. As we navigate the challenges posed by technological gaps, it is essential to recognize the interconnectedness of these segments. A holistic approach that considers the entire value chain—from fiber manufacturing to garment production is paramount. Collaborative efforts between governments, industry players, and educational institutions can facilitate skills development, technology transfer, and research initiatives to bridge these gaps effectively.

In conclusion, addressing technological gaps is not a mere technical requirement; it is a strategic imperative for the creative textile industry in South Asian countries. It is a commitment to preserving cultural heritage, fostering economic growth, and leading the way in sustainable practices. As the loom of progress continues to weave the threads of innovation, the creative textile industry in India, Pakistan, and Bangladesh stands poised for a future where tradition and modernity coalesce harmoniously, creating a vibrant and sustainable legacy for generations to come.

References

1. Gulhane S, Turukmane R (2017) Effect of make in India on textile sector. J Text Eng Fash Technol 3(1):551–555
2. Memon JA, Aziz A, Qayyum M (2020) The rise and fall of Pakistan's textile industry: an analytical view. Eur J Bus Manag 12(12):136–142
3. Rana AW, Ejaz A, Shikoh SH (2020) Cotton crop: a situational analysis of Pakistan. Intl Food Policy Res Inst
4. Abbas S, Halog A (2021) Analysis of Pakistani textile industry: Recommendations towards circular and sustainable production. Circ Econ: Assess Case Stud 77–111
5. Rumi MH et al (2021) Future tradeoff under fourth industrial revolution in Bangladesh: a study on RMG sector. Int J Soc Polit Econ Res 8(1):198–215
6. Khan FD (2011) Preserving the heritage: a case study of handicrafts of Sindh (Pakistan)
7. Nazeer N, Rasiah R, Furuoka F (2021) Technology transfer, technological capability, absorptive capacity and firm performance: An investigation of the textile and clothing firms in Pakistan. Malays J Econ Stud 58(1):99–124
8. Rahman KM, Chowdhury EH (2020) Growth trajectory and developmental impact of ready-made garments industry in Bangladesh. Bangladesh's economic and social progress: from a basket case to a development model, 2020, pp 267–297
9. Mian MEU (2020) A study on competitiveness of ready-made garments for export-led economic growth in Bangladesh: issues and challenges
10. Morelli CJ (2022) Regulating the post-independence textile trade: Anglo-Indian tariff negotiations from independence to the multi-fibre arrangement. Indian Business in the Twentieth Century. Routledge, pp 38–51
11. Rahman S (2023) 15 Bangladesh Garment industry. The Routledge history of fashion and dress, 1800 to the Present, 2023, p 286

12. Mridha J et al (2022) Analysis of the competitiveness of Bangladesh RMG sector using porter's diamond model
13. Das S, Hasan MR, Das D (2024) Who is the next China? Comparative advantage analysis from top ten apparel exporting nations. Compet Rev: Int Bus J
14. Swazan IS, Das D (2022) Bangladesh's emergence as a ready-made garment export leader: an examination of the competitive advantages of the garment industry. Int J Glob Bus Compet 17(2):162–174
15. Habib MM, Sabah S, Debnath D (2022) The scenario of fire safety through accord in Bangladeshi garments industry. Open J Soc Sci 10(4):149–165
16. Islam MM, Ali MS (2018) Industrial research advances of jute in Bangladesh. Int J Agric Biosyst Eng 3(1):1–9
17. Sikder MAR (2019) A review of textile industry in Bangladesh. Int J Adv Multidiscip Res 6(3):9–14
18. Acharya PA, Samani VS (2017) Review study on detailing traditional motifs and techniques of Bandhani and its changes of Saurashtra region. Int J Home Sci 3(1):377–379
19. Sethi MH, Shen L (2020) Ceramic art: the Kashi tiles of Sindh. Ceram Art Percept 115:130–135
20. Haque MS, Nahar N, Sayem SM (2021) Industrial water management and sustainability: development of SIWP tool for textile industries of Bangladesh. Water Resour Ind 25:100145
21. Kim YR, Kwon MJ (2021) Traditional embroidery in India based on Bourdieu's cultural theory. Fash Text Res J 23(6):758–769
22. ur Rehman A, et al (2019) Productivity improvement through time study approach: a case study from an apparel manufacturing industry of Pakistan. Procedia Manuf 39:1447–1454
23. Dixit P, Lal RC (2019) A critical analysis of Indian textile industry: an insight into inclusive growth and social responsibility. Russ J Agric Socio-Econ Sci 88(4):53–61
24. Dehghani M, Goyal P (2022) Design and development of textile fabrics using 3D printing technology. ECS Trans 107(1):19313
25. Kumar D (2022) Developing regional value chains in the textile and garment sector of South Asia
26. Noor A et al (2022) A review of artificial intelligence applications in apparel industry. J Text Inst 113(3):505–514
27. Naqvi SA et al (2020) Implementation of sustainable practices in textile processing Mills of Lahore, Pakistan. Pol J Environ Stud 29(1)
28. Jahanzaeb S (2023) Plan for successfully moving toward Industry 4.0: case of a Pakistani Textile Firm
29. Naseem S, Kashif L, Mahmood S (2020) Role of computer programming in processing and management of textile industry—a review. Asia Pac J Emerg Mark 4:79–92

Basis of Circular Economy in Creative Industry

Rabbia Asif

Abstract This chapter deals with the significance of the concept of Circularity and technological advancements illustrating the future of creative industries. Humanity is confronting the biggest threat of climate change and resource pressure while new technologies are sweeping through the society and business sphere. This scenario demonstrates the transformation which is happening at a rapid pace than ever before affecting our lives, our society even our world. Countries around the globe are striving to unlock the potential of human creativity and creative industries by integrating them with technology in order to achieve sustainable development goals. The traditional practices and creative industries of South Asia is closely linked with economic growth of the region and enjoys a distinctive position globally. Recognition of this sector with strategic framework by the governments and relevant stakeholders both at local and international level will lead to a creative ecosystem and sustainable future.

Keywords Creativity · Industry · Economy · Sustainability · Technology

1 Introduction

"There is no doubt that creativity is the most important human resource of all. without creativity there would be no progress and we would be forever repeating the same patterns."

—*Edward De Bono*

Leonardo Di Vinci, Isaac newton (1643–1727), Charles Darwin (1809–1882), Thomas Edison (1847–1931), Marie Curie (1867–1934), Albert Einstein (1879–1955), Walt Disney (1901–1966), Steve Jobs (1955–2011) are all among the most celebrated creative genius from history who shaped the cultural zeitgeist through their innovation and curious nature.

R. Asif (✉)
Pakistan Institute of Fashion & Design, Lahore, Pakistan
e-mail: rabbia.asif@pifd.edu.pk

H. Jamshaid and A. Dad (eds.), *Creative Textile Industry*, SDGs and Textiles,
https://doi.org/10.1007/978-981-97-3802-1_7

Creativity encompasses various aspects of our lives and has been a focus of psychologist, neuro scientists, engineers, performers, educators, writers, artists and even the policy makers. It delineates;

- The tendency to generate new ideas and novel concepts
- A mental process in which something valuable is created
- A potential of being able to think outside the box
- Trying out new things; more daring, experiential and playful.
- A potential and innate skill that can be learned and strengthened through practice in order to unleash our genius for living more fulfilling life.

The idea of creative industry has its roots in history from eighteenth century. With the advancement of technology and the adoption of various media forms in 1990s, the idea of creative industry evolved from the terms "Creative Arts" (individual talent wherever humans think, do and make things) and "Cultural Industries" (mass scale, links the human attributes with large scale organized enterprise) [1].

Creative industries are not like old-fashioned industries, not confined to one single industry. It varies geographically, depending on the circumstances and local heritage. People apply their creative skills or talent while creating something or devoted to product innovation. It is vast in extent and scope ranging from media, arts, architecture, sports, fashion, cultural activities, education to health, finance, steel, airline or automobile industry. The UNESCO Framework For Cultural Statistics 2009 illustrates cultural domains including (i). Cultural and Natural Heritage, (ii). Performance and Celebrations, (iii). Visual Arts and Crafts, (iv). Books and Press, (v). Audio-Visual and Interactive Media, (vi). Design and Creative Services, (vii). Tourism, Sports and Recreation https://uis.unesco.org/sites/default/files/documents/unesco-framework-for-cultural-statistics-2009-en_0.pdf [22].

Creative industries possess a global relevance and play a significant role as a major contributor to society and economic growth. In 2002, The Mayor's Commission on the Creative Industries was established by London Development Agency as a program of research and consultation. John Hawkins responded to it and mentioned economy, management, innovation and intellectual property as the four major factors and challenges for creative industries. (i) The growth in the creative economy is a key driver responsible for the change in contemporary economy globally. The businesses who are selling their ideas are majorly leading this economy which includes media and entertainment industry. (ii) A distinctive management style is notable in creative industry depending on the nature of inputs and outputs. Most important management factors include role of individual in relation to the organization, creative entrepreneurs, role of producer, joint venture or temporary company, just-in-time person, post-employment job, finance, network office and portfolio management (deals or hints). (iii) Creativity is non-linear and illogical personal expression whereas innovation deals with calculated novelty. (iv) Intellectual property (IP) is the core and currency of creative economy since we are living in the world where everything has to be quantified, where data is most significant [2]. Andy Pratt, a world's leading expert on cultural economy stated that "because these industries historically have not been taken seriously, the people not bother to collect data but if you want to survive in a political world, you need number" [3].

2 Creative Industries; Now and Next

Creative attitude is required ever since towards navigating turbulent times of transformation and dwindling certainties impacting human conditions. Things are changing at rapid pace. What does this change or transformation mean for creative industries and for those linked to these industries in long term?

Creativity and innovation go hand in hand, at times seem like similar concepts but creativity centered around original thoughts and idea generation whereas innovation turns that idea into a viable solution or in other words it is the application of creative ideas. Per Espen Stokne, a renowned psychologist, economist, TED global speaker, professor and director for the Center for Sustainability and Energy at BI Norwegian Business School in Oslo, Norway mentioned in his book "Tomorrow's Economy: A Guide to Creating Healthy Green Growth" that history has witnessed huge waves and different phases of innovation with a massive impact on society and economy. These waves of innovation "come, swell, boom, break and recede" [4]. Historians have categorized five major waves of innovation since the industrial revolution;

HISTORY OF INNOVATION	
1.1760–1830	Mechanization; mechanical factories, capital owners, period of exponential growth and innovation
2.1830–1900	Steel, Steam and Railways; changed trade, cities, coasts, settlements and agricultural productions
3.1900–1970	Industry; mass productions, consumer goods, electricity, power pumps, petroleum chemistry, highways, cars & trucks
4.1945–1990	Electronic, Television and Aviation; telecommunication and news-broadcasting networks, fighter jets, mass aviation, airports
5.1985–2000	Digital and Internet Wave; internet connectivity, mobile phones, tablets, information technology, websites, mass manufacturing, oil wave, Google, Amazon, Facebook, Twitter
6.2015–2060	Green wave; resource management, smart solutions, solar energy, technology innovation, e-platforms, clouding, robotics, AI, AR/VR, IoT

Source Author

The creative industries and creative economy are major growing sectors in the present time impacting our social, economic, political and cultural life along with the environment.

The Creative Economy Outlook 2022 by UNCTAD (https://unctad.org/system/files/official-document/ditctsce2022d1_en.pdf) [21] defines the creative industries and creative economy as follows:

2.1 Creative Industries

- are the cycles of creation, production and distribution of goods and services that use creativity and intellectual capital as primary inputs;
- constitute a set of knowledge-based activities, focused on but not limited to culture and heritage, potentially generating revenues from trade and intellectual property rights;
- comprise tangible products and intangible intellectual or artistic services with creative content, economic value and market objectives; and
- stand at the crossroads of the artisan, services and industrial sectors;
- are at the heart of the creative economy.

2.2 The Creative Economy

- is an evolving concept based on creative assets potentially generating economic growth and development;
- fosters income generation, job creation and export earnings while promoting social inclusion, cultural diversity and human development;
- embraces economic, cultural and social aspects interacting with technology, intellectual property and tourism objectives;
- is a set of knowledge-based economic activities with a development dimension and cross-cutting linkages at macro- and micro-levels to the overall economy;
- is a feasible development option calling for innovative, multidisciplinary policy responses and inter-ministerial action [5].

Technology has enabled the creative industries to push the creative boundaries while integrating artistic expressions, skills and talents with Augmented Reality and Virtual Reality, unlocking the potential for the advancements never seen before. On the other hand, creative industries are also transformative in nature in terms of job creation, earnings and income generation. It builds the confidence and empowering people while stimulating their creativity for more sustainable growth and livelihood.

Creative industries take the central stage when it comes to creative solutions for the global challenges; the COVID-19 pandemic, unemployment and poverty, increase in population, war and regional conflicts, environmental crisis, climate change and global warming, natural resource crisis and many other.

The significance of creative economy toward the development of sustainability was highlighted in the United Nations General Assembly resolution 74/198 and recognized the way it aligns with social, economic and environmental dimensions. It also highlighted the increase in revenues for different countries generated by creative goods and creative services in recent years [5].

3 Circularity; a Call for Action

The rise of developed human societies and production systems have interrupted the balance of this planet and natural systems. The way we produce and consume natural resources and different products is highly unsustainable and this wasteful attitude has polluted our world. The excessive consumption of energy, water, metals and other resources harms the nature and is threatening for the survival of future generations. This waste has become immensely a challenge of interconnected and challenging world.

In September 2015, United Nations decided on new 17 Global Sustainable Goals with 169 associated targets which are closely integrated and indivisible. The focus is to fully implement the agenda till 2030. It is a transformative step to shift the world on to a collective journey leading to a resilient path of sustainable development; the social, economic and environmental dimensions:

"53. The future of humanity and of our planet lies in our hands. It lies also in the hands of today's younger generation who will pass the torch to future generations. We have mapped the road to sustainable development; it will be for all of us to ensure that the journey is successful and its gains irreversible."

"55. The SDGs and targets are integrated and indivisible, global in nature and universally applicable, taking into account different national realities, capacities and levels of development and respecting national policies and priorities."

"56. Each country faces specific challenges to achieve sustainable development, and we underscore the special challenges facing the most vulnerable countries and, in particular, African countries, least developed countries, landlocked developing countries and small island developing States, as well as the specific challenges facing the middle-income countries. Countries in situations of conflict also need special attention." [6].

The setbacks of linear economy "take, make, dispose" and the wasting of planet's valuable resources has given way to the concept of circularity. The multidimensional impacts of linear model include negative effects on our environment, increased extraction of natural resources and raw materials (which are depleting with time) for the production of new products, impacts of extraction processes, degradation of land, rise of sea levels and global temperatures, droughts, loss of bio-diversity, scarcity of fresh water and lastly an encouragement to the culture of increased consumption in which the end-of-life value of materials are not recognized and the waste is eventually ended up in landfills. Unfortunately, we are running out of landfills also with significant environment degradation. This alarming situation has given rise to the urgent need of protecting our natural resources and to adopt a systemic approach globally towards circularity.

The concept of circularity is apparent throughout the nature with a balanced ecosystem which is in accordance to human and other lifeforms. The model of circular economy is still in infancy but has become a lead puppet for the future of business, industries and economies. It emphasizes the need to re-think the system, the attitude towards extraction, production, consumption and repurposing the materials in order to ensure the reduced emissions of greenhouse gases and efficiently keeping the material or resources in circulation as long and as much as possible; *close the loop.*

The model of circular economy is both *regenerative* and *restorative.* By regenerative it means that the processes in the system are able to regenerate and renew the energy or materials that they consume. By restorative it means that processes use less energy than would be needed to start from scratch. Both significantly aim to phase out the waste from the ideation stage, building systems that value the resources, re-use the products and materials, develop new retail models, boosting markets for used products with consumer awareness and ownership to make right choice.

Various European countries have adopted First Circular Economy Action Plan by European Commission in 2015 which demonstrates the urgent need to re-design the way economy works from root to tip. The focus is on eco-design, to adopt regenerative models in order to reduce resource consumption, reduce carbon footprints and material use rate. It is a future oriented agenda for a transformational change, for a cleaner and competitive co- creations involving consumers, economic actors and different organizations in the society.

Transitioning the way, method and processes of manufacturing a product is challenging. Many sectors especially the textile and fashion industry are embedded with older models or ways of doing things, most of the waste is either burned or sent to landfills. In an inter- connected globalized economic system, trade deals are containing circularity elements so EU's action plan has a massive impact on global supply chain.

The opportunities linked with circularity have to be easy to implement and retain the responsibility and ownership of the material. We as a society, as consumer also need a cultural and attitudinal shift towards functional value of the materials and products instead of a habit of buying new and more every time. There is a need to develop a lifestyle to eliminate waste, prefer buying sustainable products over cheap, encourage wardrobes repairing and restyling. These measures will bring social, economic, environmental and health benefits to everyone in long term.

3.1 Circular Design

From the society in which we live, the food we eat, clothes we wear, the buildings in which we reside, the organizations in which we work, interfaces that we use daily, design is the most crucial and major determinant. Design in the linear world starts by digging the ground, ends by throwing stuff away and in the middle, we produce something valuable, we as consumer buy that thing; a sense of something from nothing and filling up the planet.

Since the third industrial revolution linear thinking dominated the world led to prosperity and growth along with many unsustainable problems the world is facing today. Design thinking is an important point for circular economy. The sense of meaning comes from designing and the role of design comes in almost in every point in the cycle. Designers are the problem solvers. They have a responsibility to design a better world, to think about design ideas that have highest possible performance impact.

"Circular Design is a mission-led practice of design based on the principles of the circular economy and systems thinking" (Ellen MacArthur Foundation, n.d.). Designers' role is very significant in the selection of appropriate materials with lowest environmental impact or footprints and how the life of material can be extended according to the circular model.

They need to think of the raw materials, the outcome and the entire spectrum between these two poles; combining the craft of design and technical element in a sustainable way. Not every material is suitable for a circular product because of the chemical concerns that may be hazardous or polluting the environment. So circular thinking requires a holistic set of solutions in order to fight with over-consumption, increased productions and wasteful practices making this world a better place to live.

Visualizing the circular economy, the Butterfly Diagram by Ellen MacArthur Foundation (https://www.ellenmacarthurfoundation.org/circular-economy-diagram) illustrates the concept and continuous flow of materials into two cycles; the Biological Cycle and the Technical Cycle. In order to regenerate nature, nutrients from the biodegradables need to be returned back to earth in biological cycle. Whereas through re-use, repair, re-manufacturing and re-cycling the materials and products should be in circulation in technical cycle.

The circular design thinking approach allows us to explore ways to re-design our surroundings and our world a way to build creative confidence. We need to re-think and re- train ourselves. The new generations of designers are much more fluid with tools and technology. They are intuitive with the idea of circularity, the elements and principles of circularity. Since they are or will be grown up with it so they can bring it in work more naturally.

4 Technology; Harnessing the Creative Potential

Technology is dictating and illustrating the forms of change, hopes and fears across a range of things, a kind of a mix; fear for ourselves, our people and communities, the relation of human with the world. What do we think, what we want in our society and in the world actually reflect the way we use technology every day, challenging the idea of being human.

Digital technology has driven the fourth industrial revolution which is fundamentally transforming the creative sector also referred as Creative Industry 4.0. Klaus Schwab the founder and executive chairman of World Economic Forum has pointed out that the fourth industrial revolution is evolving at an exponential pace distinct in its scope and impact. It is characterized by a range of new technologies blurring and fusing the spheres of physical, digital and biological worlds transforming the entire systems of governance, management and production [7]. 3D printing, augmented reality (AR), virtual reality (VR), artificial intelligence (AI), internet of things (IoT), cloud computing, block chain, drones and automation systems have opened new horizons and unlimited creative opportunities. These technologies have enabled new avenues for the production, distribution and consumption of creative goods and services with a global reach.

The trend towards digitalization, e-commerce and B2B platforms has been a major shift specially after Covid-19 pandemic both in developed and developing countries. E- commerce is noteworthy for the growth of handicrafts and traditional practices of artisans especially in developing countries like Pakistan, India and Bangladesh. From the finest muslin of Bengal, yarn dyed ikat of Gujrat, Odissa and Andhra Pradesh to Pashmina shawls of Kashmir various traditional handwoven textiles have been produced and traded for centuries to different parts of the world but now these communities have embraced the use of e-commerce and digital platforms for keeping their legacy alive. The adoption of social media including WhatsApp, Instagram, Facebook has provided them an opportunity to promote their businesses and has enabled them to sell their products directly to the consumer. Jaypore (https://www.jaypore.com), Etsy (http://www.etsy.in), Amazon-Karigar (https://www.aboutamazon.in/news/small-business/how-amazon-karigar-is-reshaping-the-future-of-handmade-products) are among the online craft-based market places connecting artisanal communities, independent artists and textile designers from all over India, delivering exceptional value to the members and customers. Yali, an e-commerce space launched by The Registry of Sarees (TRS) a Bangaluru based curatorial platform in collaboration with non-profit Shrenis Trust brought back the Kodiyala sari unique to Padmashali weavers (https://www.thevoiceoffashion.com/fabric-of-india/the-south-india-edit/the-kodiyala-weave-born-again-4222). Shrenis has provided digital opportunities to the artisans including the use of WhatsApp, hosting online Q&A sessions, workshops and showcasing the products. E-commerce and digital platforms have the potential to offer new and expanded business opportunities, market research, international export, content quality and traffic management for entrepreneurs and local communities. However, the creative industry and communities need adequate support and infrastructure to harness opportunities and benefits of hybrid markets, to set up e-retails and to connect with e-retailers. Digital piracy and plagiarism should be called off in order to ensure the self-sustaining economy [8].

UNCTAD's eTrade Readiness Assessments initiated in 2017 aims to provide a diagnostic and policy advice fostering an enabling environment and building e-commerce ecosystems both in developing and least developed countries (LDCs). In 2021 the number of eT beneficiaries reached 29 taking follow-ups and actions at varying speed. This program is supporting and helping countries to harness the potential of e-commerce and embracing digital trade in order to meet the Sustainable Development Goals. It has also tailored the need to increase the dialogue among public–private sector, partner's commitments and coordination at ministry-government level [5, 23].

On one side it has changed the lifestyle and shopping patterns of consumers while on the other the hand people have turned towards more online streaming services and subscriptions. Digital streaming of music, movies, gaming and drones have an impressive impact on media productions. Streaming platforms such as Netflix, Spotify, YouTube have become leading channels and major revenue generators. Improved advertisements, television productions, film-making, aerial photography, and journalism are significant of drone technology. Entrepreneurs, small businesses and many independent artists suffer and face multiple challenges due to already concentrated digital streaming market.

Jonathan Tilley, one of the senior experts in McKinsey's Southern California office highlighted two key ingredients of the 4th Industrial Revolution including automation systems and programming flexibility, important in defining Creative Industry 4.0. He shed light on the modern production concept embedded with automation and robots. Since the automation systems are becoming more smarter and cheaper, they are becoming more adaptable by employing robots as an alternative to human labor. Cost effective and economically viable productions are possible due to flexible programming and niche tasks are being performed by the robots in different companies working with significant product variety and small batch sizes. For example, in one of the FANUC plants in Oshino, Japan, one of the largest robotic companies in the world, industrial robots are being employed to produce industrial robots which are supervised by a team of only four workers per shift. In 2013, Canon camera making company phased out human labor replacing by robots. In the next few years, it is expected that there would be a fundamental change in the nature of tasks to be performed by robots. Emerging technologies are expected to achieve higher levels of precisions by simplifying robot programming's, improving voice-recognition technology or the possibility for giving them verbal instructions. Companies need to adopt a systematic and holistic approach in order to capture full value provided by these opportunities and align them for present and future business needs [9].

AI is likely part of the future. It is everywhere in everything with a potential to broaden the creative palette; in our mobile phones, social media feeds, video games, banking systems, surveillance etc. In the past few years, the field of AI has been expanded to numerous creative fields and is continued to grow. It is emerging as a collaborator, as a new tool instead of a competitor in many creative industries. John McCarthy who coined the term "Artificial Intelligence" in 1956 defined it as "it is the science and engineering of making intelligent machines, especially intelligent computer programs. It is related to the similar task of using computers to understand human intelligence but AI does not have to confine itself to methods that biologically observable" [10].

These machines are artificially incorporated with human like intelligence to perform multiple tasks. These are built by using complex algorithm and mathematical functions.

Generalized learning, reason and problem solving; these three capabilities make the machines artificially intelligent. In short, they have the ability to adapt, reason and provide solutions (Fig. 1).

The difference between AI, machine learning and deep learning is that machine learning is a technique to achieve AI whereas deep learning is a subset of machine learning. Machine learning provides a machine with a capability to learn from data and experiences through algorithms. Deep learning does this learning inspired by the

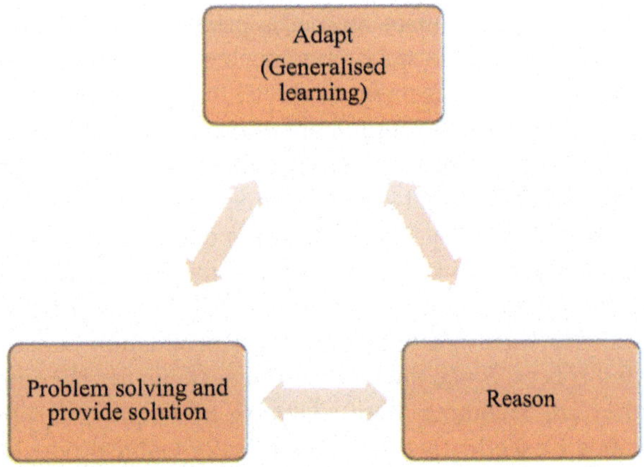

Fig. 1 Capabilities that make the machine artificially intelligent. *Source* Author

human brain (neural network) so data and patterns can be better perceived. Whereas GenAI is a subset of DL which generates high quality images, texts, codes, sounds and videos (Fig. 2).

The advancements in AI have posed questions for the future of creative industry plus the role of practitioners in creative processes. AI is pushing the boundaries with remarkable progress in generating visuals based on the provided prompts but still current systems need human interactions in terms of aesthetic values, artistic or creative visions and decision making. In performing arts, AI can generate content, scripts, dialogues, story-telling, valuable insight for character development but lacking the depth of human emotions, interpretation, personal experiences and spontaneity. In music industry it can replicate melodies, musical compositions but still cannot match the personalized touch or style of composers. AI can enhance the picture quality, can generate realistic visuals, aesthetically pleasing but still lack the human eye and empathy.

Experts have envisioned and made mind-boggling predictions regarding the future of AI, robotics and how they will impact the economic and employment scenarios in

AI Technologies		
Machine Learning (ML) capability to learn from data and experiences through alogrithms. For example to identify groups, correlations or values.	**Deep Learning (DL)** a subset of machine learning, develop connections inspired by human brain (neural network) so data / pattern can better be perceived	**Generative AI (GenAI),** a subset of deep learning, generates high quality contents including codes, images, texts, audioes and videoes

Fig. 2 Different AI technologies. *Source* Author

coming years permeating vast areas of our lives. Ray Kurzweil who is a well-known author, inventor, computer scientist, futurist and director of engineering at Google envisioned a possible future of human–machine synthesis in SXSW conference held in Austin, Texas. He expounded the idea of singularity which according to him has already begun and by the year 2045 robots will be as smart as humans [11]. Elon Musk's vision of mind-controlled technologies may be a long way off. He anticipated that human mind and body will be enhanced by AI implants which will make us partly CYBORGS [12].

With more precise and advanced technologies, industrial robots are becoming more artificially intelligent and agile, these "cobots" have the potential to automatically adjust their movements. Advanced sensors, computing power and vision technology has enabled them to assist and work side by side with human labor rather than replacing them; collaborative machines [13]. The debate about AI and creative industry is ongoing with a recognition of human value and potential of AI, there is a need to address implications at ethical, legal and societal level.

Human skills are one of the basic elements to pursue the digital and technological transformation of creative sector. The potential it holds calls for development of human capacities, investments in education systems, access and training to use different digital platforms, providing the right set of skills and digital competencies leading to an increase in employment opportunities. Generic digital competencies include literacy, design skills, entrepreneurship and marketing are essential for the professionals for digital creativity.

Technological disruption is often assumed to lead to job losses, but such anxieties are frequently over blown.

5 South Asia; a Creative Hub

The region comprising Pakistan, India and Bangladesh is dynamic with a significant appetite for research, innovation and collaboration in education, industry and creative sector. Creative industries specially the intergenerational textile practices served as a fertile ground for creativity and innovation that evolved across Indian Sub-Continent from time immemorial till present day. The geographical location, rich history and material culture shaped the creative and conventional practices which are deeply integrated in the social, cultural and economic dynamics of the region. In fact, these practices bridged the connection across various cultures and desired around the world.

From the deserts of Sindh and Baluchistan to North-West Frontier, from Rann of Kutch and Coromandel Coast to the villages of Bangladesh, India truly is the most original and prolific center of textile production in the world. The use of gold and silver metal in lustrous and luxurious jeweled textiles (Zari or Zardozi), Kinkhab Brocade, finest qualities of Malmal and Jamdanis, delicate Chikan-Kari work, complex yet sophisticated woven Kashmiri Shawls, unsurpassed designs of Ikat (*double ikat woven patola),* finely printed Odhnis and Ajarakh, figurative and floral Kalamkari work, Painted Roghan fabrics, finely woven Khes (*blankets*), distinct designs and rich colors of folk Embroidered Costumes, Appliqued Quilts, Wall Hangings decorated with beads and sequences, Hand-Painted Shrine Cloths, Embroidered Cradle Cloths, ply-split handmade Camel Girths, multi-sensorial textiles and furnishings to accentuate the interior spaces are the reflection of unique legacy and artistry of the Indian Sub-continent [14].

Cultural and creative industries in South Asia are synonymous with originality and dynamism deeply rooted in traditions, tangible and intangible heritage with a blend of global trends. It possesses great respect, openness and innovation for the benefits of economy and society. There is an increase in the overall level of creative literacy in society in recent years. The adoption of technology with creative life is playing a major part for the promotion of international profile, embedding sustainable activities at local and provincial level.

Textile and garment industry in South Asia gained remarkable growth in the recent years along with record breaking textile exports. After China, Bangladesh stands as the second largest garment producing country in the region accounts for four-fifth of the total country's exports. Whereas Indian home textile exports increased by 10.2% between the years 2017 and 2021, making 10% of global exports [15]. Pakistan ranks as the eighth largest exporter of textile commodities in Asia contributing 8.5% to the country's GDP but the year 2022 was uncertain and challenging for the textile industry. After the setbacks of Covid-19 pandemic, global recession in the year 2022, Ukraine war, devastating floods, dwindling exchange reserves, precarious energy and its increasing cost situation have affected textile exports in Pakistan. Despite the heavy investments made by textile players in the last 2–3 years, the global economic slowdown, shortage of spare parts, chemicals and other items have hampered the growth of industry. This situation of energy crisis and inflation has called for the attention and the need for immediate solutions [16].

Net Zero Pakistan is a collaboration at the national level between different sector experts, institutions and companies in order to accelerate the sustainability initiatives and deliver the goal of achieving zero carbon by 2050. It brings global partners to support signatories in their ambition of net-zero;

- The government roadmap outlines the pathways for signatories
- Signatories commit, undertake, measure, eliminate and report greenhouse gases
- Capacity building through support program, discussions, workshops for all stakeholders
- A unified branding system in order to align and demonstrate sustainability credentials to buyers, suppliers and investors
- Centralized financing facility
- Fact based support for policy to enable net-zero transition

Several textile industries have become the signatories of Net-Zero including AGI DENIM, Al-Karam, Al-Rahim, Artistic Milliners, Diamond Fabrics, Gohar Textiles, Gul Ahmed, Interloop, International Textiles, Liberty Mills, Mahmood Group, Sapphire Textile Mills, Midas Safety, Tetra Pak, US Apparel and Textiles,

Fig. 3 Recyclable Handbags by *Sapphire*. *Source* Author

Yunus Textile Mills, Siddiq Sons Group and many others. Carbon pathway is a flagship program by Net-Zero seeks to restore a natural eco-system in order to reduce emissions and retore bio-diversity. (Pakistan Enviornment Trust, n.d.) (Fig. 3).

With due course of time, multiple factors altered the nature of creative industries, with time consuming nature of handcrafted objects, expenses for material procurement, and lack of demand being the major determinants. In order to cater to the changing demands of the society, contemporary artisans, practitioners and designers are using their resources to revive the age- old techniques and struggling to keep their creative legacies alive. The experimentation with a range of indigenous techniques, materials and processes have resulted in respecting the tradition, an openness to the changing social needs and tastes along with a nurtured relationship between makers, designers and consumers. During 1980s textile and fashion designers, clothing brands, industries and even the artists have started focusing on traditional attires, ornamentations and handcrafted techniques. Artisans and designers started experimentations with drapes, cut lines, different materials and techniques envisioned new possibilities in their workshops and studios. Thus, the eco-system of makers, producers, designers and consumers generates different forms of livelihood and provides opportunities for boundless creative innovation in contemporary times.

Indian Famous designers, **Abraham and Thakore** reinterpreted the imagery of sari into a contemporary and office look using handwoven shirt fabric in Houndstooth Pattern originated in Scotland (https://collections.vam.ac.uk/item/O1273328/sari-and-matching-abraham-thakore/sari-and-matching-abraham-thakore/). The black and white created a contrast with bright yellow panels whereas instead of tweed fabric, silk added softness to it. The full sleeved dress shirt with same patterns, contrasting colors and material exemplifies creative innovation.

Bibi Russell, a Bangladesh based clothing and homeware brand aims at promoting traditional designs, local textiles and crafts with makers at its core. **Generation** is one of the leading ready to wear brand of Pakistan making efforts to embrace sustainability towards a greener future. It is working with a vision to sell craft-based work to be worn by current and future generations which represent the unique beauty of different towns, villages and the skill passed from one generation to another. It includes different embroideries, ajrak, applique and so on. Among many initiatives the biggest is "reGENerate" utilizing surplus and waste fabrics for apparel, accessories and samples. The sorting center of generation sorts out the fabrics and other materials which can be used again. The policy of no signal plastic has been adopted in the production line and have introduced custom made bags (Generation, n.d.) (Fig. 4).

The role of entrepreneurs, practitioners, professionals, initiatives by different NGOs and government at national and provincial level, employment and business opportunities are significant in providing energy to the creative sector. Female artisans/crafts-makers have a leading role in creative industries. They are benefiting by this sector as public and private initiatives have been taken to raise their skills, providing them sustainable livelihood, creating more awareness of international markets and opportunities to make connection with wider audience. **Dastkari Haat Samiti** is a non-profit Indian organization with members from all the states. It functions independently following the democratic procedures, voluntary help of

Fig. 4 Textile accessories by *Generation. Source* Author

its supporters and member artisans (https://dastkarihaat.com). **Aarong**, Bangladesh based ethical brand offering high quality lifestyle and fashion products. This lifestyle retail chain aims at promoting artisans and offering unique market for handicraft with the blend of tradition and contemporary styles. Through on-line networks, fairs and international trading the brand is providing livelihood to more than 70,000 artisans with an opportunity to retain age-old crafts [25].

An ethical fashion brand **Inaaya** by Naushaba Brohi is focusing on Sindhi handicrafts by harnessing the power of village entrepreneurship and empowering the women there. It has become the face of an artisanal fashion in the global market narrating the journey of Pakistan from craft to couture. Whether it is an accessory or an attire, each piece is unique having the beauty of imperfection actually providing back self-confidence and self-reliance to the makers [24].

Polly and Other Stories, a Pakistan based inspirational lifestyle brand with a mission to revamp and elevate the craft / cottage industry to meet the demands of modern marketplace (https://www.pollyandotherstories.com/pages/meet-the-makers). The brand engages with the artisans from rural areas specially females changing the narrative around makers and creative businesses. It provides support through design, product development and training interventions.

Vastrakala, created by Jean Francois Lesage, Patrick Savouret, Malavika Shivkumar and Sandeep Rao in 1993 became a subsidiary of Lesage Paris in 2014 located on the outskirts of Chennai, Tamil Nadu. It is one of the leading hand embroidery companies contributing to the global textile and fashion industry. Around 200 embroiders are registered and full time employed at the workshop (https://vastrakala.in/index.php/the-workshop/). The embroiders as the descendants of master craftsmen who served Nawabs of Deccan over centuries are the real inheritors of traditional techniques and passing down their legacy from one generation to the next. The embroiders undergo trainings to enhance and upgrade their skills with master trainers from Paris aiming to regain pride and confidence in the future of their craft (Vastrakala, n.d.).

Different artists and practitioners have adopted the textile mediums and have prompted new ways of inquiry, exploration of broader themes and creative imagination in contemporary times. **Ayesha Khalid** a leading Pakistan based contemporary artist incorporated Indo-Persian traditions and rich symbolism in her work. Her work investigates the socio-political aspects, the notions of spirituality and conflict, gender role and women's right through an eastern lens. The artists' diverse practices comprise of mix media small- and large-scale paintings and range of textile mediums including embroidery techniques along with gold pleated pins (https://artspectacleasia.com/aisha-khalid-at-aicon-new-york/).

Monika Correa—an Indian textile artist interpreted the tradition technique of weaving with modern sensibility. "The technical ingenuity that Monika began to experiment with, i.e. removing the reed during the process of weaving, enabled her to switch, within the body of the piece itself, from the structured order of conventional weaving to the relative chaos of the unreeded areas. This technique became an important determinant of her aesthetic improvisations. She inculcated into her work elements of a minimalist abridgement of nature's forms and, above all, a kinetic quality—moving branches, flowing water, mingling shadows, whistling wind and a magical play of shadow and light."- Jyotindra Jain [18].

New generation of entrepreneurs are emerging in the field of media, visual arts, film and television, performance, providing creative service and design solutions yet more encouragement is needed towards knowledge exchange, greater collaborations and interactions at both national and international level. Performance industry is still betwixt and between within the cultural value system. It includes theatrical activities, drama, dance, music, story-telling, mimicry, puppetry both in traditional and contemporary forms. This industry is still attracting audiences, increasingly in demand and offering a platform for a variety of creative activities, services and products. The trans-disciplinary collaborations and interactions between literature, film, theatre and television are becoming the global norms.

Indian entertainment industry, film and media emerged as strongest domain for the potential growth. Digital transformation, advanced technology and IT are the key drivers behind this entertainment revolution. Indian television is ranked third largest in the world whereas the film industry produces more than 800 films annually in different languages including Hindi, Bengali, Tamil, Punjabi, Telugu and Malayalam [19].

There is an emergent need of embedding relevancy of higher education curriculum with current and contemporary creative practices. The absence of data recording and policy consistency in comparison to international market can hinder the progress of this sector, identity development and future potential. The recognition and image of creative industry outcome is still a crucial challenge for sustainable international collaboration and trade.

Besides all the social, political and economic challenges the lived reality of South Asia should not be under-estimated. The creative industry as a combination of rich cultural heritage with contemporary expression enjoys a prominent as well as emergent position globally.

6 Conclusion

Predicting the future where the world will go considering what is happening at this time gives a clear picture that ***our choices matter*** significantly. Technology is advancing at a rapid pace so the machines can perform many of the logical and analytical functions but creative skills and human imagination or a sensitive soul cannot be programmed into them. A kind of mix, a meaningful connection and a balance of human creativity with technology will be the constituent part of businesses, organizations, our society and communities. For sustainable human development, countries around the world are striving to shape the world by recognizing the potential of creative industries which is vast and unlocking with the passage of time. The concept of creative economy and the sector of creative industries vary country wise according to the regional and cultural diversity, economic structure and statistics. It is "a multitude of different local trajectories" [19], affirming to

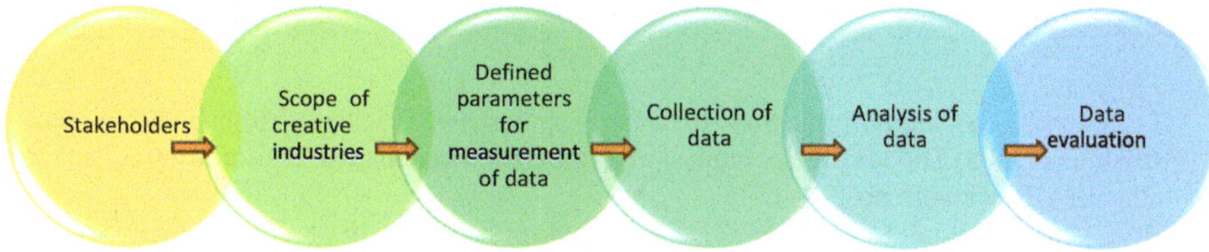

Fig. 5 General framework for measuring creative sector performance. *Source* Author

the distinctive identities. The key drivers for these industries are human creativity and innovation at both individual and group level. In order to re-design and get the maximum output from the creative sector considering the future scenarios it must be harmonized with existing classifications of national and international organizations including UNESCO, WIPO (World Intellectual Property Organization), IDB (Inter-American Development Bank), UNCTAD (United Nations Conference on Trade and Development).

It is essential to identify, map and measure the international trade contribution and export revenues generated by the creative sector. It will benefit the policy makers in providing insight, setting realistic objectives, parameters and evidences for re-designing, re- planning and informing stakeholders while highlighting the critical areas which need to be focused. Policies in developing countries can identify the challenges and barriers for entrepreneurs, micro, small and medium enterprises. Communication, lack of access to technology, digital skills, limited access to payment methods, cost of logistics and absence of copyright regime in rural areas are among the major challenges. Involvement of national statistical officers, authorities and organizations is significant in compiling the data, recording the trade revenues, GDP and employment rate (Fig. 5).

Data regarding the marginalized communities, involvement of females, youth and informal workers is of utmost importance for the policy makers with transparency and confidentiality. International organizations can provide support, resources and guidelines to developing countries by contributing with knowledge sharing, joint ventures and capacity building. In order to provide protection to consumers, digital data, small businesses, digital markets, cybersecurity and e-payments fostering e-commerce, governments must develop and adopt regulations and legislative frameworks.

It can fulfill the trade gap due to lack of data measuring and recording between developed and developing countries especially in the case of creative services. Because with the advancement of technology, there is a major shift in trade of tangible products / goods to delivering creative services. This sector has generated more revenues in recent years and has the potential for accelerating income generation, employment, trade and investments.

References

1. Hartley J (2005) Creative industries. In: Creative industries. Blackwell Publishing Ltd., pp 4–6
2. Howkins J (2005) The Mayor's commission on the creative industries. In: Hartley J (ed) Creative industries. Blackwell. pp 120–124
3. City, University of London (2022) Accessed https://www.research-in- focus.city.ac.uk/creative-force/
4. The MIT Press Reader (2021) Accessed https://thereader.mitpress.mit.edu/what-250- years-of-innovation-history-reveals-about-our-green-future/
5. United Nations (2022) Creative economy outlook 2022. United Nations Publications, New York. Accessed https://unctad.org/system/files/official-document/ditctsce2022d1_en.pdf
6. United Nations (2015) Transforming our world: the 2030 agenda for sustainable development
7. Schwab K (2016) The fourth industrial revolution: what it means, how to respond. Retrieved from World Economic Forum: https://www.weforum.org/agenda/2016/01/the-fourth-industrial-revolution-what-it-means-and-how-to-respond/
8. Ranavaade VP (2021) Sustainable craft design systems for handloom weavers. In: TEXTILE, cloth and culture, pp 399–409
9. Tilley J (2017) McKinsey & Company. Accessed https://www.mckinsey.com/capabilities/operations/our-insights/automation-robotics- and-the-factory-of-the-future
10. McCarthy J (2007) 1 what is artificial intelligence?
11. Futurism (n.d.) Futurism. Accessed https://futurism.com/kurzweil-claims-that-the- singularity-will-happen-by-2045
12. Financial Times (2019) Accessed https://www.ft.com/content/eec3bfb2-aa09-11e9- b6ee-3cdf3174eb89
13. Hollinger P (2016) Financial times. Accessed https://www.ft.com/content/6d5d609e-02e2-11e6-af1d-c47326021344
14. John Gillow NB (2008) Indian textiles. James & Hudson Ltd., London
15. Ray T (2024) International finance corporation. Accessed https://www.ifc.org/en/stories/2024/a-green-dress-code-for-south-asias-textile-and-apparel-industry
16. Baqai A (2022) Was a year of uncertainity and full of challenges for the textile industry. Pak Text J 7. Accessed https://sdgs.un.org/sites/default/files/publications/21252030%20Agenda%20for%20Sustainable%20Development%20web.pdf
17. Generation (n.d.) Accessed https://generation.com.pk/pages/press
18. MASH (n.d.) Accessed https://www.mashindia.com/art-and-the-code-of-textile-part-i/
19. UNESCO (2013) Creative economy report
20. British Council (n.d.). Accessed https://www.britishcouncil.org/sites/default/files/cultural_and_creative_industries_in_pakistan_0.pdf
21. UNCTAD (n.d.) Accessed https://unctad.org/system/files/official-document/ditctsce2022d1_en.pdf
22. UNESCO (n.d.) The 2009 UNESCO framework for cultural statistics (FCS). UNESCO. Accessed https://uis.unesco.org/sites/default/files/documents/unesco-framework-for-cultural-statistics-2009-en_0.pdf
23. United Nations (2022) Fast-tracking eTrade readiness assessment implementation
24. NOT JUST A LABEL (NJAL) (n.d.) Accessed https://www.notjustalabel.com/inaaya-0
25. Aarong (n.d.) Accessed https://www.aarong.com/about-aarong

Applications of Artificial Intelligence in Textiles and Fashion

Hassan Anwer, Mumtaz Ali, and Hafsa Jamshaid

Abstract Artificial intelligence (AI) and its fields are strong tools for enhancing performance and productivity in textiles and fashion. Starting from design concepts to future forecasting of upcoming trends, there exist numerous applications of AI tools. Here, we summarized the working principles related to AI tools commonly used for textiles and fashion. The application of these tools for fault detection, quality assessment, and color assessment of textiles is discussed. Similarly, uses of AI tools in concept and design, production, logistics and distribution, retail, and fashion forecasting are explained. AI enhances production speed, decreases labor cost, improves the quality of production, and provides a variety of options to users, therefore its uses are currently emerging. Few of these applications are commercially used, however some of these tools are at the research stage. For example, there are numerous papers related to the application of AI for fabric fault detection, however, there are several limitations to be addressed for practical use. This chapter also provides future directions and a detailed summary of currently used AI tools in the textile and fashion industry.

Keywords AI · Fashion retail · Textile production · Creative design · Trend forecasting

H. Anwer
Department of Environmental Engineering, National University of Sciences and Technology, H-12, Islamabad 44000, Pakistan

M. Ali (✉)
Department of Chemical Engineering, Hanyang University, 222 Wangsimni-ro, Seoul 04763, Republic of Korea
e-mail: Mumtaz@hanyang.ac.kr

H. Jamshaid
School of Engineering and Technology, National Textile University, Sehikhupura Road, Faisalabad 3760, Pakistan

H. Jamshaid and A. Dad (eds.), *Creative Textile Industry*, SDGs and Textiles,
https://doi.org/10.1007/978-981-97-3802-1_8

1 Introduction of Artificial Intelligence for Textiles

Artificial intelligence (AI) has improved the quality of life in nearly all aspects of life, including textiles and design. Textiles are porous and fibrous structures used in clothing and upholstery and are known for their flexibility, comfort, and suitable strength. Textiles are the second basic need of human beings; hence their consumption constitutes an evergreen thriving field. Textiles were mainly used for shelter and protection in the stone ages. With the passage of time, textiles have evolved to represent culture, social status, and gender. The advancement in textiles is progressing at a fast pace, where computers are playing a role in the design, manufacture, testing, and analysis of textiles' properties. In this digital era, the role of AI has become a cornerstone in the development of advanced textiles, from fiber to garment level. This chapter will review the latest trends of AI applications in textiles. In 2023, 140 articles were published on the application of AI in textiles and overall, 7000 citations were recorded for the publications related to AI. Amongst all countries, 32% of the publications of AI in textiles are from China, followed by the USA and Iran contributing 10% and 8%, respectively, in this research area. From the abstract analysis of the previous research, the most common research areas observed were fiber, yarn properties, fiber, fabric defects, texture, and fashion [1]. Scheme 1 summarizes the AI application in the whole fashion industry. Based on these results, the research of AI on textiles can be classified into:

(1) AI for textile structures

Fiber structure, density, diameter, and morphology were the top keywords of the literature. Yarn count, yarn hairiness, yarn diameter, yarn image, yarn strength, and yarn twist were the parameters studied related to yarn structure. Fabric-related parameters mostly studied were related to fabric surface, holes, porosity, fabric density, wrinkle, surface roughness, stiffness, drape, and fabric appearance. Another area of textile structures includes non-woven and composite structures, where mechanical and thermal properties prediction is of primary interest. Keywords used in fiber-reinforced composites-related clusters include failure analysis, geometry, damage image processing, image processing algorithm, and digital techniques of detection. Comparatively, yarn analysis was related to hairiness analysis, yarn diameter imaging, variation analysis, color indexing, and impurity detection. Particularly, the artificial neural network technique is most commonly used for yarn analysis. In digital printing, the color matching and its replication is also achieved via AI [2, 3].

(2) AI for textile inspection

Conventional manual inspection systems used in textiles for product quality evaluation suffer from issues such as time consumption and low accuracy. AI can accelerate the inspection process time and accuracy for quality control while using simple image processing techniques. Extensive research on AI applications in fabric inspection has been conducted utilizing computer vision and image analysis, construction parameters, weave pattern detection, defects analysis, and impurity detection. The

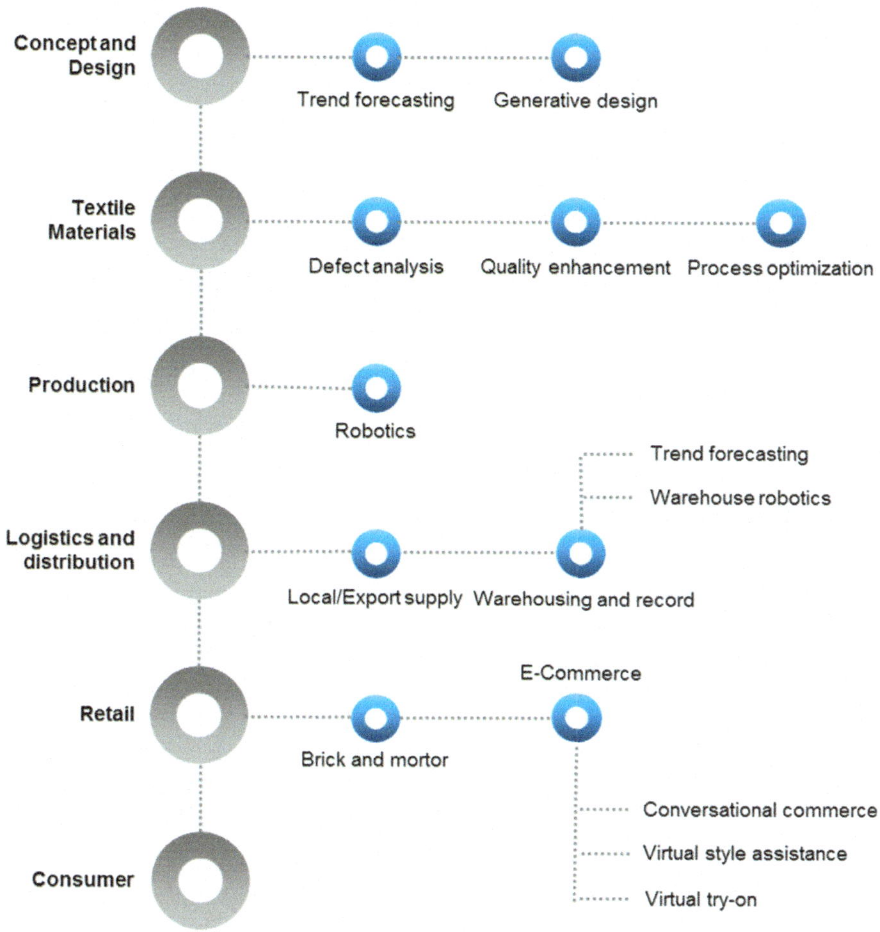

Scheme 1 Schematic summary of applications of AI in the fashion and textile industry (Self drawn figure)

automation of the process is carried out using high-speed cameras, which can also be installed on the weaving machine for the early detection and instant repair of the fabric faults. The process utilizes computer tools such as design methodology, pattern recognition, support vector, convolutional neural network, feature extraction, image segmentation, deep learning, and genetic algorithm. Similar techniques are also used for the detection and removal of fiber contamination, fiber color detection, blending analysis, and impurity detection in the yarn manufacturing process [4].

(3) AI for fashion

AI is a key tool in predicting the ongoing trends and future of fashion industry. Augmented reality-based style assistance is an important application of AI, which enables virtual try-ons for visualizing how textiles and accessories will look on customers without physically trying on the outfits. This enhances the customer's interest in online shopping by providing an immersive platform to make better decisions for clothing selection. Virtual fashion shows have also gained interest allowing viewers to experience fashion shows from the comfort of their homes. AI applications in smart textiles include yarn sensors for detecting body posture, monitoring work environment, time and production prediction, and person in real-time. In textile production, AI is being used for automatic laying and cutting pattern adjustment. Automatic laying systems with high accuracy are used for the placement of fabric layers and compact arrangements of patterns, which reduces fabric waste during the development of intricate and complex clothing fashion items. In addition to automatic laying systems, AI has been employed for garment shape prediction, garment draping, image acquisition, body scanning followed by pattern making, and prototype designing [5].

2 AI Techniques

AI is a branch of computer science dedicated to the study of computation requiring intelligence if performed by human beings. Specifically, the term "intelligence" refers to the computational activity related to achieving goals.

AI-based systems have become popular due to their ability to perform the functions that require human beings with better accuracy and speed. It is a broad field, which covers a broad spectrum of human–machine interactions. Machine learning is a branch of AI that uses experience to sort out patterns in a given data set. A computer is taught and trained by using previous examples, which enables it to recognize patterns, unlike conventional programming with specific rules.

2.1 Machine Learning

Based on algorithms and type of learning, there are three main types of machine learning systems, including (as shown in Fig. 1):

- Supervised learning
- Unsupervised learning
- Reinforcement learning

The choice of the method being used will depend on the problem being solved and data set under consideration.

Most machine learning processes utilize supervised learning, where data is given with known correct results (Fig. 2a). There exists a relationship between inputs (X) and outputs (Y), which is sorted out using machine learning. By adjusting the algorithm, the learning function will give the fundamental equation forming a mutual

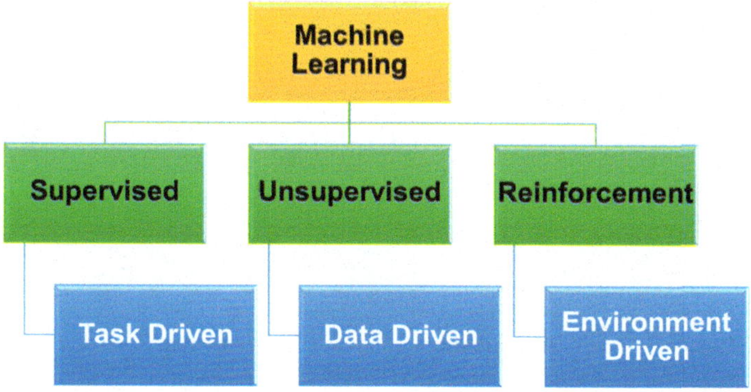

Fig. 1 Types of machine learning (self drawn figure)

relationship between both input (X) and output (Y) (Y = f(X)). It is not necessary to give extensive values of X or Y, however, based on a suitable data set, we can predict an overall trend or relationship between input and output. Therefore, the same function can be used for future decision-making processes. For example, for a new input (X), we can precisely guess the output (Y), by using the model generated by the supervised system. The rules were supervised based on the previous data, which enabled the learning function [2].

In unsupervised learning, the input of data is known, whereas the corresponding output is unidentified. The objective of this method is to know the trend of data and the underlying structure from the data distribution. Unsupervised learning has no predefined output; therefore, it is called data-driven technique. The algorithm is left by their alterations to find out the data structure, which allows them to resolve the problems without having any idea of the results (Fig. 2b). Data clustering to understand the data trend is one of the methods of unsupervised machine learning [6].

Reinforcement learning is based on trial-and-error method, which resembles a baby in a learning phase. The machine comprehends information at first and based on the self-driven trials and results, it will decide whether the output is either good or not. If the decision is unfavorable, the model will explore alternative options. Conversely, if the output has required features, the algorithm will consider it a good decision, prompting further progression (Fig. 2c). In this way, the model learns through an iterative process resulting in an automatic improvement in the machine's performance over time. The alternate option or action command is generated by the agent, which is applied on the environment. According to the response, the agent will decide the next action. Positive reward reinforces the action and negative reward suppresses the action, finally resulting in a suitable outcome.

AI is applied in several areas of life including robotics, artificial vision, artificial neural networks, comprehension, and expert systems. AI has particularly assisted in solving problems of textile sector. The contribution of expert system, artificial neural networks, and fuzzy logic for advancing textiles will be discussed in the next section.

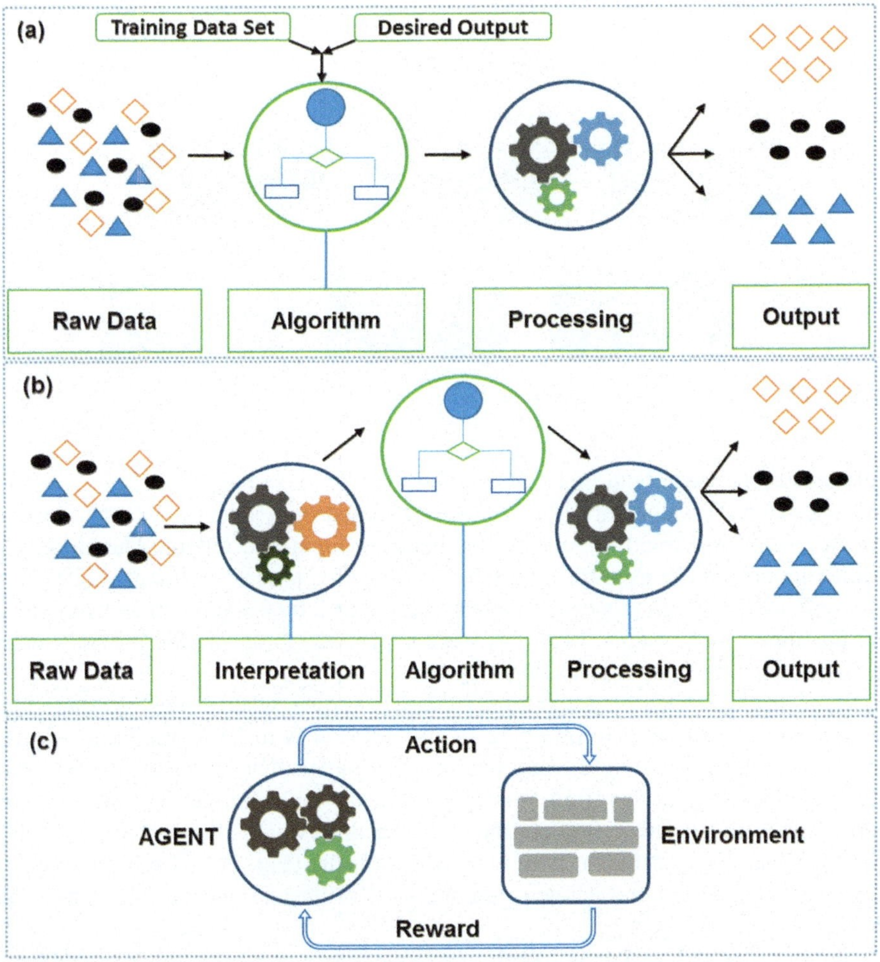

Fig. 2 Schematic illustration showing the types of machine learning, including, **a** supervised learning, **b** unsupervised learning, and **c** reinforcement learning (self drawn figure)

2.2 *Fuzzy Logic*

Neuro-fuzzy logic is the most common AI technique used in the textile industry. This is a hybrid of neural network techniques and fuzzy techniques. A neuro-fuzzy system is an embedded hybrid system consisting of the combination of two well-understood modeling methods, such as artificial neural networks (ANN) and fuzzy logic (FL). At present, these hybrid systems are of great interest, as they bring the benefits from FL and ANN, thereby eliminating the disadvantages of the individual techniques. Due to the better performance of the hybrid technique, it is commonly used in continuous industrial processes. For example, the neuro-fuzzy systems are used for the prediction of fabric and garment drape, based on the cotton fiber properties [3]. In textile dyeing, color matching is challenging due to variations in lighting conditions and dye-bath batches. These variations in color perceptions can be minimized by creating color-matching systems using FL. Similarly, the mechanical and texture of knitted fabrics can also be well forecasted, however yarn parameters prediction using this system is still a challenge.

2.3 Artificial Neural Networks

Artificial neural networks are formed by algorithms developed for a specific purpose, which exactly resembles the working of human brain. For example, a system obtains knowledge from various learning steps which is stored in synaptic weights to memorize the knowledge. Connections between the neurons are called synapses, which are assigned different values known as synaptic weights. Therefore, the name artificial neural network is given to this system, as it resembles the network of neurons, serving as a single processing unit [2].

Artificial neural network was used to reduce the yarn defects and fabric defects. For example, the inspection of fabric was conducted with high-speed camera and the images of the fabric were used as input of the artificial neural network system. This image acquisition system compared the input images with the saved reference standard images, where the mismatch is considered as defect. Image analysis and comparison can suppress the dimensional defects, while accurately comparing new features with existing old features.

2.4 Expert Systems

This technology is a recent derivative of artificial intelligence, which has extensive applications based on its own knowledge. Scientist utilize the knowledge of expert system to generate innovative solutions for advancing technological development in various fields. An expert system can mimic human experts by using computational systems to explore the underlying reasoning of specific knowledge.

Main components of expert systems include:

(a) Knowledge base: it consists of the facts, principles, and scientific theories related to the knowledge of experts in the area of our interest.
(b) Development team: it is based on experts, who can translate the set of knowledge described previously into the production standards.
(c) Development environment: it is composed of user-friendly development environment, also referred as AI shells. These shells can rapidly and accurately generate interface screen for user, hence capturing the knowledge base to manage the search methods in the standard base, as shown in Fig. 6.

The system of specialists can also be used in several ways in the textiles and fashion industry. Expert systems are being used for coloring validation after application to fabric. The task carried out by the system of specialists in the management of production is assisted by the expert system AI. A similar mechanism is used in the expert system developed by Sandoz AZ's company, under the commercial name of Wooly. This system is used to approximate the color in dyeing. Wide range of color recipe mixtures have already been added to the system for different fibers and fabric types. By using its built-in library, this system can predict and suggest the pattern and method of coloring. Wooly system is coupled with the dyeing machines to accurately develop the dyeing recipe, enhanced dyeing method, compatible dyes, and to match the color system. The expert systems are known for their utility in textile supply chain management. Expert systems consider numerous factors, such as product demand, lead time, and production capacity to provide valuable information for decision-making with respect to inventory management, order fulfillment, and logistics.

3 AI for Textiles

3.1 AI for Yarn Analysis

This section will cover the aspects related to yarn analysis using the AI techniques discussed above. Image processing is majorly used for identifying yarn variations; artificial neural networks are used for predicting the mechanical properties based on input variables; and finally, expert systems are used to assist decision making in various production and design processes.

Both automatic and manual inspection systems are used in the textile industry. In the case of yarn, these inspection systems assist in evaluating the yarn quality regarding thick-thin places, trash content, and yarn hairiness. Manual inspection poses severe issues regarding the complexity of testing and time wasted in the inspection. Moreover, the accuracy of the manual system is compromised due to human fatigue and attention lapses, which enhances the possibility of omitting faults at various points. Therefore, the reliability of such a manual system can be challenged.

Automatic inspection systems using AI can make these processes fast and easy. The quality of automatic inspection is superior to that of manual inspection systems. Furthermore, AI systems can be installed on the machine, thus low quality yarn formation can be avoided at the spot. This way, the overall quality of production is enhanced. Artificial neural networks can also be applied to data to analyze the correlation of material or yarn type and defect types.

The image processing technique uses a high-speed camera to take the images at high resolution, which are then converted to mathematical values. Differences in the mathematical values during actual production signify a variation in yarn, thereby detecting the yarn defect. This is a reliable solution to overcome the problems of manual visual analysis. The ongoing research on this technology is constantly improving the quality and reliability of AI based inspection systems. Recently, a system was developed to analyze the neps of cotton yarn and extract fiber orientation data in the entangled network. The mathematical equations derived from the neps data are analyzed for defect examination using probabilistic neural networks (PNN) [7]. The test was performed using k-fold cross-validation technique, which was conducted via image processing of the PNN system. The results demonstrated that the AI classification system was valid for the detection of neps. Researchers are extending the application of these methods to predict the mechanical properties of yarns based on the types of neps in yarn [7].

Uster Technologies AG has previously invested in yarn analysis systems capable of predicting the yarn hairiness and yarn slubs (thick places). However, the accuracy of these systems was compromised due to low-quality imaging systems and constant yarn movement. This might cause improper imaging on the high-speed spinning machines. In comparison to Uster system, a model with higher accuracy was developed, which was known as YSQ®. Some of the preliminary algorithms were used to obtain better results, in which the reference images of the AI system were taken at high speed [8]. The YSQ system consists of an image acquisition and analysis system, with a microscope attached via USB camera. With a combination of suitable hardware and software systems, the precise variations of yarns can be detected. The software of this system was developed by IMAQ vision, which played a key role in image capturing and analysis. Also, this system works better in the presence of monochromatic light, resulting in a better contrast of image and associated yarn variation. The application of this system has now been extended to analyze fiber twist, yarn orientation, number of yarns in a compound yarn structure, and compaction of fibers in yarn.

3.2 AI for Fibers and Fiber Spinning

Artificial neural networks are used for the analysis of various problems in the textile industry. A variety of fiber processing variables can critically affect the properties of the product. Fiber properties are non-linear, and it is challenging to predict the properties with high accuracy. It is essential to understand the performance of a product as a function of fiber parameters as system inputs. The dynamic nature of artificial neural networks is advantageous for the prediction of fiber properties and associated textile products. For example, the fiber properties formed via co-polymers can be predicted using an artificial neural network. Because co-polymer contains two or more types of repeating units, therefore the content of each monomer can be altered to achieve the required performance. The properties of fibers depending on the content of monomer are related by using the AI system. Two types of fibers can also be sorted by using artificial neural networks, as shown in Fig. 3. For instance, Merino wool and Mohair wool were identified using artificial neural networks [9].

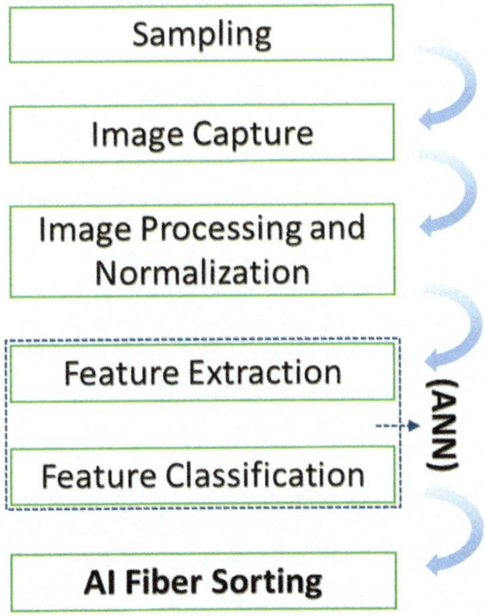

Fig. 3 Algorithm of artificial neural network for fiber classification (self drawn figure)

The AI model uses the six characteristic scales via image processing and the remaining data is extracted using unsupervised networks. Based on the complexity of the images and the accuracy of the final model, the classification of fibers is made. Further colors can also be used in the classification and identification process. With slight modifications, the same system can also be used for the detection and classification of cotton fibers, where fineness and fiber length are used for the differentiation of cotton fiber types.

Figure 4 shows the artificial neural network for cotton fiber and yarn, to approximate the mechanical properties of yarn. Based on this model, the elastic properties of yarn were predicted from the fiber parameters obtained from testing. This way cost was reduced for the hit and trials in yarn spinning. Also, production time can be reduced by linear programming along with an ANN algorithm [10].

3.3 AI for Fabric Analysis

Inspection is important to ensure the quality of the products. Real-time on-spot inspection can also reduce materials and energy wastage. It is estimated that 60% of textile waste occurs due to defects. If these challenges are overcome, the quality can be significantly enhanced while reducing waste and saving valuable resources. Currently, fabric production is slow, therefore real-time inspection during weaving is a challenging task. Therefore, the inspection process is conducted after weaving by

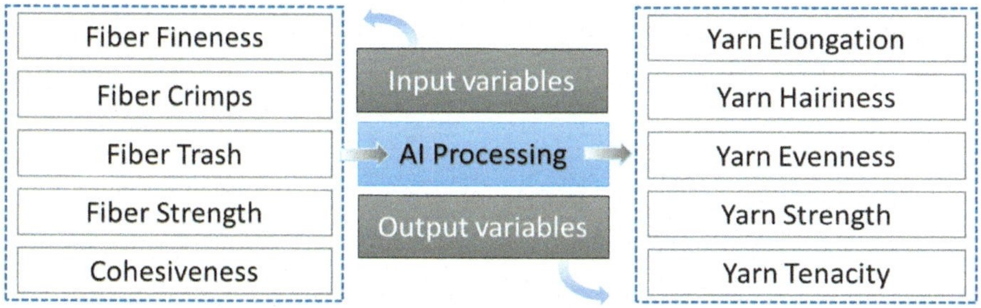

Fig. 4 Algorithm of artificial neural network for predicting yarn properties based on fiber characteristics (self drawn figure)

laborers. Which being a labor-intensive and time-consuming process poses a limit on production.

The fabric structure can be analyzed from its appearance and look; for example, woven, knitted, or braided fabric structures each has its texture and repeating unit in the image. By using the difference in texture via visual imaging can allow the use of online inspection. This can reduce labor fatigue and human errors in fault detection. For the purpose, high speeds cameras are used to record and analyze the fabric structure and based on observation, out-put is "OK" or "NOT OK", for defect-free or with defective fabric, respectively.

In the start, vision system is trained to understand the patterns of the fabric. It is also made to understand the tolerable limits of defects in yarn color and texture. This teaching time can be varied depending on the type of image quality and imaging zoom power. It may take a few weeks in training, after which it can manage the detection of faults online. This process is specifically important for knitted fabrics, where defects can only be properly visualized under stretch conditions. However, there are unlimited possible structures and color combinations in fabric, therefore it is a difficult task to manage all fabrics using a single code. On the other hand, woven fabric with plain weave is a common product used in the weaving industry, therefore most of the literature is related to woven fabric analysis [11].

Feed forward neural network is used for the detection of fabric faults. This system relies on converting the fabric image to greyscale and then allotting it mathematical numbers, based on the depth of its grey scale. If there is a fabric defect, the grey scale of the fabric mesh is changed, thus the arrangement of each consecutive pixel is changed. This way each segment has its color pattern on a grey scale, which can only be disturbed in case of a defect.

Depth of grey scale is linked with the information vector, which is altered by the difference in the fabric texture. Initially, the texture of a broader region is measured, with which newly developed fabric texture is matched. The neighboring pixels forming the information vector are responsible for the detection of defects [11]. One of the examples of difference in imaging due to defects in woven fabric is shown in Fig. 5.

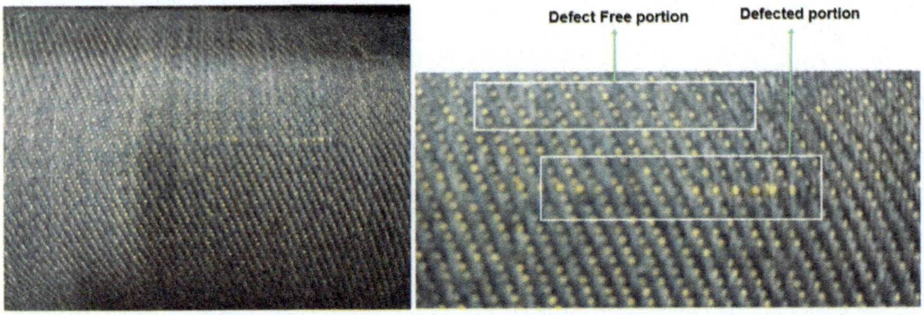

Fig. 5 Analysis of defect-free and defective fabric, based on the color difference or texture difference (Image courtesy Weaving Lab, National Textile University, Faisalabad, Pakistan)

3.4 AI for Color Matching

Color of the textile product can be within the acceptable limit or sometimes it is not acceptable because of "very light" or "very blue", or "very red". These types of analysis are done based on visual analysis or using color measuring tools. The color of the fabric is matched with the required color and color difference is indexed based on the tolerance limit. Numerical descriptors are used to accurately depict the color [12]. One of the simplest AI methods is based on comparing the developed sample with a reference sample and marking pass or fail (P/F). This way accurate matching of the color can be achieved with the color pallet and human errors can be avoided. Recently, an advanced AI system for color matching has been developed based on artificial neural networks and "Kubelka–Munk Theory (K-M)". The underlying mechanism behind color is based on partial absorption and partial dispersion of light by the specific-colored fabric. The dispersed light gives color, whereas the absorbed light generates heat in the fabric [13].

In order to estimate the color of the colored fiber's blend, ANN was used. Similarly, the difference of fiber type can also be taken into consideration, while determining its final color. Specifically, the accuracy of K-M theory is significantly improved after the use of AI, as consistency of test measurement is possible with these tools. Therefore, currently a system of K-M and ANN combined is considered more precise in predicting the reflectance values [14].

The K-M function for color analysis is:

$$\mathrm{F(R)} = (1 - \mathrm{R})^2/2\mathrm{R} = \mathrm{k/s} = \mathrm{Ac/s} \quad (1)$$

where:

R = Reflectance, s = Scattering factor, A = Absorbance, C = Concentration of absorbing species, and k = absorption coefficient.

The color is evaluated by comparing the color spaces under standard illuminations. The hybrid method K-M + ANN is conducted in the following three steps:

(1) Defining the equivalent tissue substrate via a simplified model.

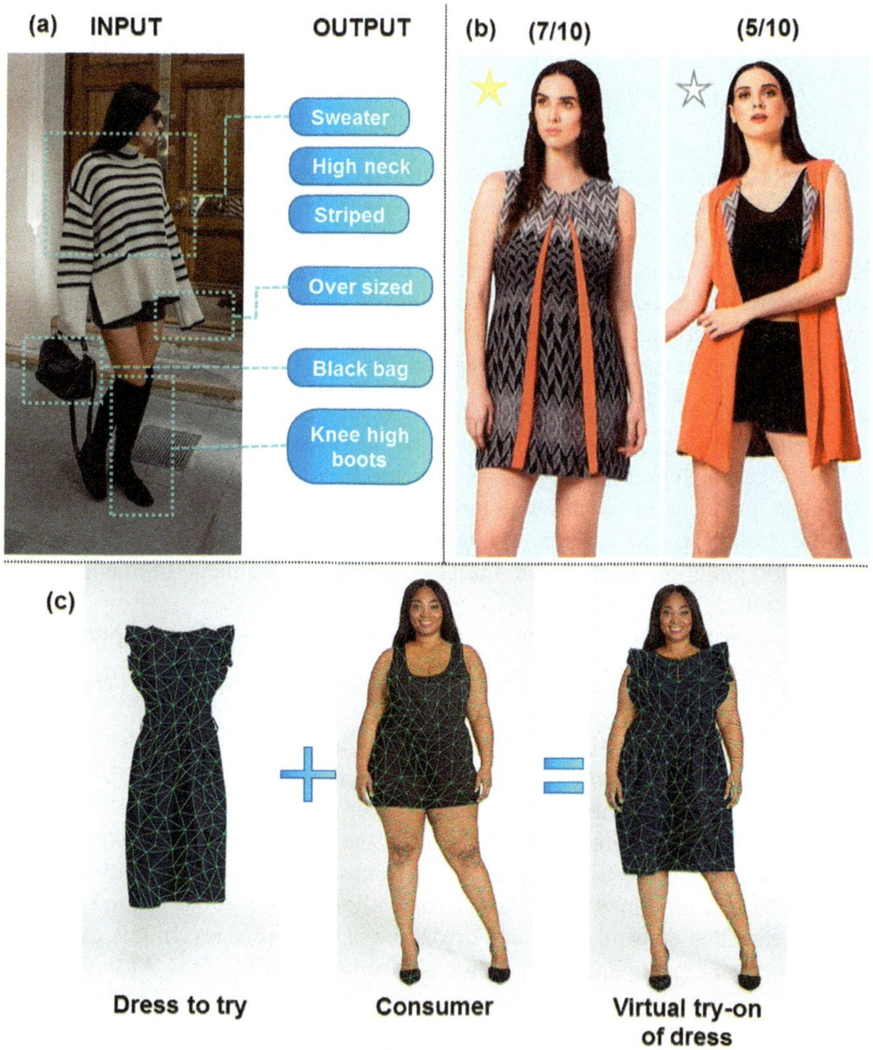

Fig. 6 Applications of AI for **a** analyzing the online images, **b** virtual style assistant, and **c** virtual try-on (Self drawn figure)

(2) Training ANN system to interpret the differences of recipe in the K-S ratio of the substrate: which is then used to guess the optical response of the tissue mixture. The combined effect of both steps 1 and 2 will result in a hybrid system of K-M + ANN.

(3) Construct a practical graph from the hybrid system that can be used to simply calculate the color combinations without repeating the whole lengthy process.

4 Applications of AI in Fashion:

4.1 Concept and Design

The fashion projects are developed based on the concept and design, thus making it a critical process to impact the whole fashion chain. There are different approaches for the concept and creative design, for example, one of the aspects could be innovation-driven design. A commonly used approach for the design is human-centered design, which is intended to meet the specific requirements of people. Designers can integrate emerging technologies into the design to ensure business success. For this process, the following steps are taken into consideration:

(1) Empathize: We must design the product while considering ourselves a user.
(2) Define: Analyze the needs and problems associated with the existing system. Based on the observations, the insights are discussed.
(3) Ideate: It includes the thinking process to evaluate the challenge and devising the strategy to encounter the challenge assumption.
(4) Prototyping: It involves the initial feasibility assessment of the proposed solution.
(5) Test: A successful initial trial is then extended for industrial-scale production.

It is not necessary to execute this process in the above sequence. These process steps can also overlap.

Trend is a common term used in the fashion industry, which includes the designs or colors commonly used at the time. Trend research involves understanding contemporary trends and their links with past experiences and predicting future trends. The exact prediction of upcoming prints can hype the sales of specific designs. Previously, there was a top-to-bottom influence of trends, where brands used to design the trends. Whereas bottom-up approach is currently common in the fashion industry, which means that consumer demand drives the demand process. In this case, large data from social media and other online platforms is analyzed via AI tools, which can tell us the current ongoing trends and associated future trends [15].

Data mining tools are used for the analysis of social media picture data, which can organize the information in groups and clusters. The data of consumers is collected from their social media posts, which are sorted out by the data mining tools, as shown in Fig. 6a. These tools can build patterns in large data sets, which can be used to judge the trend. The analysis of data sets can be carried out via computer vision tools to convert the data into a textual form. As shown in Fig. 6a, the minor details include the color, length, prints, and texture of the clothing. Similarly, the AI system can also extract the data from the captions and hashtags provided by the users. The system can manage more than 800,000 images at a time. After training the system for a couple of weeks, the AI can differentiate between major fashion trends, such as off-the-shoulder fashion, high rise, skirt, and tops. Classification of all these images without human assistance is possible via AI tools. The AI classification of the images allows the designers to understand the trends with respect to geographical location and to predict the evolution of clusters in the future. Similar to the data of social media posts, the AI tools can also be used on runway shots and other e-commerce images. Next-ATLAS is one of the companies for fashion forecasting.

The concept of computers replacing human beings for creative purposes is not science fiction anymore, specifically after the launch of generative designs. Using AI, a completely new artistic look for fashion can be created, like one produced by artists and designers. A new image is produced by using the given requirements and previously existing visual information. For instance, previous street-style images of specific areas will be used as input and based on generative models, a new look will be created by the AI system [16].

By using deep learning techniques, the textures and possible options for specific garment types were created (Yildirim et al. 2018). The variation in colors and shapes of the garment can be pretested using a generative adversarial network (GAN). The ultimate goal of the process was to accelerate the fashion design process. This also provides a variety of options for the designers and consumers to choose from different variations, thus saving time and providing higher possibilities of style. In addition to the AI system, GAN uses an additional layer to make the AI-generated models closer to real-life products.

4.2 Production Process

Fashion and textile are labor-intensive industries and the future of these industries will rely on solutions to reduce the workforce while maintaining high production volumes and quality with minimal defects and material wastage. Similar to the use of AI-assisted inspection systems, robotic arms are a topic of interest in the garment manufacturing industry. The robotic systems can assist in the spreading of fabric, cutting via laser arms, and sewing operations [17]. Recently, a startup: SoftWear from Atlanta, USA has developed a nearly automatic system for garment stitching. According to their claim, their robotic system can sew a single T-shirt in 2.5 min, using a computer vision system and robotics. This system has the potential to reduce labor by around 90% in cutting department. Furthermore, sewbot can also stitch other products like pillow covers, mats, mattress covers, and towels. In the future, we might see complex products being processed via sewbot. However, currently this technology is only available in the USA.

Sewts is a German startup focusing on the design of advanced robots for fabric and garment handling. For this purpose, cameras are embedded in the robotic arms, which can use computer algorithms for the proper handling of the garment. The AI system of the Sewts can approximate the flexibility and softness of a given garment or fabric piece. According to its drape, the folding process of fabric or garment is carried out. Recently, at ITMA 2023 exhibition in Italy, a robotic arm able to replace the finished cone with a new cone for weft insertion during the weaving process was displayed. This technology resembles Sewts, however, this replacement is not a frequent operation. Also, looms are equipped with an option of feeding an extra cone in the feeder, to automatically start using extra cone mounted, without wasting time. Therefore, using such expensive and sophisticated technologies might not be an area of interest for less frequent operations.

4.3 *Logistics and Distribution*

At the moment, the warehouses using AI based solutions are less than 10%, for managing their warehouses (Robotics and Automation News, 2019). Use of AI for logistics and distribution involves inventory data management, export, or local distribution [18]. It can either speed up the process or can also assist in planning the stocks via demand forecasting. Warehousing is generally considered a low profit sector; therefore, the application of advanced technologies was not considered in the past. However, 54% of the logistic companies face labor shortages, which could be resolved by using robotic systems. Material handling and sorting can be conducted by an AI system of computer vision and image analysis. AI systems assist the warehouses in the following aspects:

(1) Mapping: robots can map the surroundings and organize the data for easy sorting of the products.
(2) Navigation: Using a computer vision system, the robotic systems can navigate in the store space.
(3) Collision avoidance: AI system helps in the identification of moving objects around. The robots can manage the high-speed transport without colliding with each other.
(4) Identification and sorting: AI systems can assist in bundling similar sizes or colors of objects. Also, the scanning of bar codes can assist in the identification of stock units.

Currently, Geek+ is a new company of smart logistics, providing AI solutions for the apparel industries' warehouses. Their 200 robotic systems can arrange 6000 racks of manufacturing units for Nike brand. For Decathlon, their facility is improved to 40,000 pieces per day and for Bosideng their capacity is 3500 pieces per hour (Geek+ , 2020).

4.4 *Retail*

Retailing develops a connection between fashion brands and their consumer. It can be a direct channel, including outlets and mono-brand stores. It can also be an indirect channel where garment is sold in specialized category outlets, such as garment shops etc. Compared to these conventional methods, e-commerce and mobile commerce are emerging as potential alternatives. This includes the use of AI and online data to generate options suitable for the consumers. More than 60% of internet users make online shopping at least once. With the growing use of smartphones and internet users, e-commerce will be the next big market for fashion.

4.4.1 Conversational Commerce

In the case of conversational commerce, an AI-operated chatbot can start a conversation, where suggestions are provided based on the customer's requirement. Chatbot also provides suggestions relevant to recent fashion shows. One to one conversation is a highly involving mode to capture customers, thus the number of users of such applications is increasing at a rate of 45,000 per month.

4.4.2 Virtual Style Assistant

Personal style assistants provide better style options according to personality. Virtual-style assistant does the same task using data, natural language processing, and machine learning. It is similar to conversational commerce; however, the major difference lies in the virtual assistant we can compare the different looks. Virtual style assistant provides a rating for different looks you want to try, as shown in Fig. 6b. The rating is based on the ongoing trend and favored looks created by AI. "Style by Alexa" is an AI-based style assistant that can recommend style designs based on voice commands. Similarly, it also has an option to upload your picture and ask, "how do I look?". After analyzing the picture using computer vision, this app not only rates your looks but also suggests trending looks.

4.4.3 Virtual Try-On and Smart Mirrors

Online shopping lacks trying on the dress, therefore most consumers prefer shopping physically from stores. Similarly, online shopping has another drawback of exact size prediction via different online stores. Sometimes there can be differences in their sizing system, which may lead to a mismatch of products. Therefore, AI is used to avoid these problems and make the online shopping system much better. Inspired from the filters used for Instagram, augmented reality is used for the virtual try-on options. This allows the users to see how they will look in a specific product, without actually wearing those products. One of the most famous and commonly used virtual try-on is from Nike, which can tell shoe size. It can also detect the difference in size between both feet. For clothing, Zeekit is a start-up to provide users with a virtual fitting room, where they can see the look of dresses available online. For this virtual try-on, the user needs to upload their body shape in the form of an image, and then the AI system will generate their looks in all online catalog clothes, as shown in Fig. 6c.

Twenty-two percent increase of online sales each year has caused the closure of various brick-and-mortars recently. To attract customers to the physical selling stores,

AI-based systems can enhance the user experience using augmented reality-based smart mirrors. These mirrors can change the color of a dress worn by the user and show the suitability of other options available at the store. Smart mirrors also offer a voice detection service, where customers can interact with the AI. It also generates the data for the stores, regarding which color is mostly chosen by customers, which item is rejected mostly, and so on. MemoMi and H&M's voice interactive mirrors are available in the stores, which combine digital imaging and augmented for a better experience at stores for customers.

5 Future Aspects

The applications of AI are currently emerging. Over time, it is expected to extend the use of AI in fast fashion even more. Following future directions are essential to maintain a balance of AI influence in the fashion industry:

- Considering the sustainability, the AI systems to access the recyclability and managing the post-consumer waste could be a key area. At the moment, there is no research or applications of AI related to this topic.
- Textile laborers having extensive experience in the field of textiles must be trained to use the updated AI technologies. A highly efficient model of co-existence of human and AI-based technologies could be made possible.
- Emerging technology addiction of AI can lead to the monopoly of a few companies in the fashion industry. The use of data by AI also poses the privacy concerns of the users.
- Emerging human-free robotic system will lead to the loss of human jobs in the future. There exists a threat for millions of workers associated with the textile manufacturing and stitching industry. An alternate option for these laborers needs to be adjusted at first.
- Individual fiber level 3D geometry of yarn and fabric for simulation purposes is not available in the literature. If we can precisely model the orientation of fibers and yarns, there can be a possibility to exactly predict the deformation behavior of fabric.
- The use of AI for analytical properties of fabric or yarn could also be a useful tool in the future. Analyzing the fiber type, fiber fineness, and length can be done via AI tools, which are significantly economical and faster than the physical testing methods used currently.
- The cost of AI-based robotic systems is not affordable for the countries having textile production. Therefore, low-cost systems with high precision might extend the use of AI in a larger public.

6 Summary

Compared to other industries, the use, and applications of AI in textiles and fashion is low. Due to this, the textile and fashion sector is one of the largest hosts of labor. With the increasing use of AI in textiles, machines can replace human beings over time. Although the practices of AI are not excessive currently, extensive work is needed to make this technology viable. Another drawback of AI systems is the lack of expert labor to manage, maintain, and train the AI-based equipment. Compared to production, the application of augmented reality in virtual try-on and trend forecasting etc. is quite promising and simple to use. Similarly, the data management of warehouses and online retailing via e-commerce are in practice. Ongoing research in electronics and AI is expected to make AI technologies more efficient and market competitive very soon.

References

1. Bullon J, Arrieta AG, Encinas AH, Dios AQ (2017) Manufacturing processes in the textile industry. Expert systems for fabrics production. In: International symposium on distributed computing and artificial intelligence. Accessed https://api.semanticscholar.org/CorpusID:54021845
2. Majumdar A, Majumdar PK, Sarkar B (2005) Application of an adaptive neuro-fuzzy system for the prediction of cotton yarn strength from HVI fibre properties. J Text Inst 96(1):55–60. https://doi.org/10.1533/joti.2004.0062
3. Veit D (2012) Fuzzy logic and its application to textile technology. In: Veit TT (ed) Simulation in textile technology. Woodhead Publishing, pp 112–141. https://doi.org/10.1533/9780857097088.112
4. Kuo C-FJ, Lee C-J, Tsai C-C (2003) Using a neural network to identify fabric defects in dynamic cloth inspection. Text Res J 73(3):238–244. https://doi.org/10.1177/004051750307300307
5. Jeyaraj PR, Nadar ERS (2019) Computer vision for automatic detection and classification of fabric defect employing deep learning algorithm. Int J Cloth Sci Technol 31(4):510–521. Accessed https://api.semanticscholar.org/CorpusID:155153741
6. Koulali I, Eskil MT (2021) Unsupervised textile defect detection using convolutional neural networks. Appl Soft Comput 113:107913. https://doi.org/10.1016/j.asoc.2021.107913
7. Ghosh A, Hasnat A, Halder S, Das S (2014) A proposed system for cotton yarn defects classification using probabilistic neural network. In: International conference on recent advances and innovations in engineering (ICRAIE-2014), pp 1–6. https://doi.org/10.1109/ICRAIE.2014.6909246
8. Pereira F, Macedo A, Pinto L, Soares F, Vasconcelos R, Machado J, Carvalho V (2023) Intelligent computer vision system for analysis and characterization of yarn quality. Electronics. https://doi.org/10.3390/electronics12010236
9. Yildiz K (2020) Identification of wool and mohair fibres with texture feature extraction and deep learning. IET Image Process 14(2):348–353. https://doi.org/10.1049/iet-ipr.2019.0907
10. Ghosh A, Chatterjee P (2010) Prediction of cotton yarn properties using support vector machine. Fibers Polym 11(1):84–88. https://doi.org/10.1007/s12221-010-0084-y
11. Chandra JK, Banerjee PK, Datta AK (2010) Neural network trained morphological processing for the detection of defects –in woven fabric. J Text Inst 101(8):699–706. https://doi.org/10.1080/00405000902812735

12. Hwang JP, Kim S, Park CK (2015) Development of a color matching algorithm for digital transfer textile printing using an artificial neural network and multiple regression. Text Res J 85(10):1076–1082. https://doi.org/10.1177/0040517515569525
13. Vadood M, Haji A (2022) A hybrid artificial intelligence model to predict the color coordinates of polyester fabric dyed with madder natural dye. Expert Syst Appl 193:116514. https://doi.org/10.1016/j.eswa.2022.116514
14. Şahin C, Balcı O, Işık M, Gökenç İ (2023) Artificial neural networks approach for prediction of CIELab values for yarn after dyeing and finishing process. J Text Inst 114(9):1326–1335. https://doi.org/10.1080/00405000.2022.2124629
15. Shi M, Chussid C, Yang P, Jia M, Dyk Lewis V, Cao W (2021) The exploration of artificial intelligence application in fashion trend forecasting. Text Res J 91(19–20):2357–2386. https://doi.org/10.1177/00405175211006212
16. Liu L, Zhang H, Ji Y, Jonathan Wu QM (2019) Toward AI fashion design: an attribute-GAN model for clothing match. Neurocomputing 341:156–167. https://doi.org/10.1016/j.neucom.2019.03.011
17. Chen L, Su Z, He X, Chen X, Dong L (2022) The application of robotics and artificial intelligence in embroidery: challenges and benefits. Assem Autom 42(6):851–868. https://doi.org/10.1108/AA-07-2022-0183
18. Giri C, Jain S, Zeng X, Bruniaux P (2019) A detailed review of artificial intelligence applied in the fashion and apparel industry. IEEE Access 7:95376–95396. https://doi.org/10.1109/ACCESS.2019.2928979

The manufacturer's authorised representative in the EU is Springer Nature Customer Service Centre GmbH, Europaplatz 3, 69115 Heidelberg, Germany. If you have any concerns regarding our products, please contact ProductSafety@springernature.com

Printed and bound by CPI Group (UK) Ltd, Croydon, CR0 4YY

30/09/2026

02225308-0018